A CREATIVE STEP-BY-STEP GUIDE TO

GROWING
FUCHSIAS

A CREATIVE STEP-BY-STEP GUIDE TO

GROWING
FUCHSIAS

Author
Carol Gubler

Flower Arrangements
Jane Newdick

Photographer
Neil Sutherland

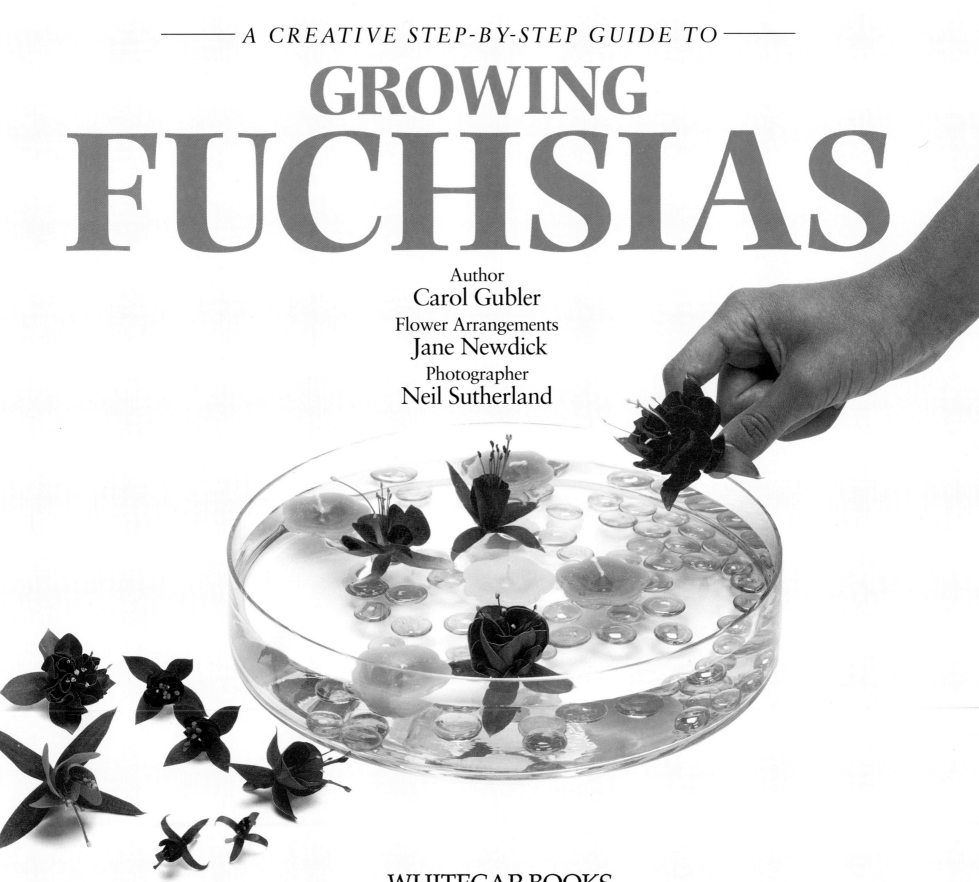

WHITECAP BOOKS

CLB 3314
This edition published 1995 by Whitecap Books Ltd
351 Lynn Avenue, North Vancouver, B.C. Canada V7J 2C4

© 1994 CLB Publishing, Godalming, Surrey, England
Printed and bound in Singapore by Tien Wah Press
ISBN 1-55110-161-0

Credits
Edited and designed: Ideas into Print
Photographs: Neil Sutherland
Photographic location: Little Brook Fuchsias, Hampshire
Typesetting: Ideas into Print and Ash Setting and Printing
Production Director: Gerald Hughes
Production: Ruth Arthur, Sally Connolly, Neil Randles

THE AUTHOR

Carol Gubler has been involved with growing fuchsias
since her early years, when she helped her father look after
his increasing collection of plants in the greenhouse. Her
interest in the subject continued through school and
university, and she obtained a BSc in Botany and Zoology
at the University of London. In 1985 she started her own
nursery and as it developed she was involved in the
introduction of a number of new varieties. She has been a
member of the British Fuchsia Society Committee since
1985 and is currently the Show Manager of the London
Show and a member of the national show committee. As
well as all this, she lectures on fuchsias, judges shows and,
most importantly, still grows fuchsias for fun!

THE FLOWER STYLIST

Jane Newdick worked for a major international magazine
company before branching out on her own to work from
her home in the countryside. She regularly contributes to a
number of magazines, as well as writing books on flower
arranging and using flowers in a variety of ways to create
unusual and beautiful decorations for the home. Her ideas
are featured on pages 96-105 of this book.

THE PHOTOGRAPHER

Neil Sutherland has more than 25 years experience in a
wide range of photographic fields, including still-life,
portraiture, reportage, natural history, cookery, landscape
and travel. His work has been published in countless books
and magazines throughout the world.

Half-title page: The vivid double blooms of Paula Jane.
Title page: Floating fuchsias form part of an exotic display.
Copyright page: Peggy King, a boldly coloured cultivar.

CONTENTS

Part One

GROWING FUCHSIAS

Fuchsias are wonderfully adaptable plants. In the wild, they flourish in mountainous areas with a damp and humid climate. In our gardens and containers, they will thrive in a range of different conditions, as long as they receive a little shade and a touch of humidity. The simple form of the natural species has been extended by hybridization into a spectacular range of delightful cultivars. Fuchsias became extremely popular during the late 1800s, and it was then that a great deal of hybridization took place, particularly in the UK and Europe. Today, fuchsias are grown and loved throughout the world. This part of the book opens with a close look at the diversity of fuchsias available today, from the traditional hardies, with their small, delicate flowers but incredibly robust growth, to the exotic, double-flowered varieties that grace countless tubs and hanging baskets during the warm months of the year. Attention then focuses on the practical aspects of growing these glorious plants to perfection, from choosing a good plant in the first place to exhibiting prized specimens at a competitive show. Of course, there is plenty of hard work to be done between these two points, and this section provides all the advice necessary to achieve perfect results, with full coverage on soils, planting techniques, feeding and watering, pruning, pests and diseases, winter care, propagation, pinching out, potting up and hybridizing a new variety. In fact, everything to enable even the newest beginner to grow fuchsias with confidence and success.

Left: Fuchsia Leverkusen growing strongly in a container.   *Right: The lovely double blooms of Ballet Girl.*

The diversity of fuchsias

Fuchsia flowers are found in all sizes, from tiny blooms just 0.25in(6mm) long to others measuring 5in(13cm) across. Flowers may be long, thin and elegant or saucer-shaped, and vary from small and dainty singles to large and exotic doubles with a mass of petals. There are flowers that drip from arching branches and flowers that look you in the eye. The range of flower colors is vast and increasing, and includes white to almost black, green and almost brown. The array of shades and subtlety is breathtaking: on the one hand, pale and demure and on the other, so gaudy that the colors almost seem to clash. In an ideal situation, fuchsias can remain in flower for twelve months, although they do appreciate a short rest at some stage!

Color is not just limited to the flowers; foliage also occurs in all imaginable shades of green. Some leaves have a matt finish, while others have a glossy sheen. Many fuchsias have variegated foliage that can include gray, cream, yellow, pink and bronze in all kinds of combinations. This variation in foliage helps to create a splash of color long before the plants are in flower.

Sample the delights of this exotic, but easy-to-grow flower and you will never be disappointed. There is something for everyone in the extraordinary family of fuchsias, so be adventurous.

Antigone, a cultivar introduced in the 1880s.

The deep violet flowers of Swanley Gem fade only slightly once opened.

Left: *Single-flowered fuchsias are delightfully simple; most have four petals or, very rarely, five. The simplicity of their flowers shows their closeness to the original species.*

Right: *A small group of fuchsias to illustrate their diversity. They are all singles, but show a wide variation in shape, size and color and allow us a glimpse into the expanding world of fuchsias.*

Alison Patricia has upward-facing flowers, which is unusual among the fuchsias.

Left: Peter Bielby is a large, flamboyant, double-flowered fuchsia. Doubles have more than eight petals, often with a number of small petals, or petalloids. Fuchsias with five to eight petals are known as semi-doubles.

Elfriede Ott is a triphylla with typical long, thin flowers.

Right: Space Shuttle is a single, but has the shape of a triphylla. It has the most unusual color combination in its flowers, with an exceptional amount of yellow. It flowers quite easily for the whole year.

Species fuchsias

Knowing something of the origin of a plant gives you many clues to growing the modern cultivars to best advantage. Fuchsia species are found in the wild in a limited number of countries, particularly in Central and South America, over a range of about 6,250 miles (10,000 km), from northern Mexico south to Argentina and Tierra del Fuego, and across the Pacific to Tahiti and New Zealand. They occur predominantly in mountainous areas or on the edge of rainforests. Only deep into the Southern Hemisphere do they venture out into the slopes and valleys. The first species were discovered in the 1800s and today there are over 100 known species, and more new ones are found every year. The plants come in all shapes and sizes, from large and treelike species, such as *F. arborescens* and *F. excorticata*, to those that creep, such as *F. procumbens*. The flowers occur in the size range 0.25-3in(6mm-7.5cm) and, like the cultivars, show a tremendous range of forms. Colors vary from white through orange, green and deep maroon, to the more conventional red and purple. The New Zealand species are especially fascinating, as both the flowers and foliage vary most widely from the accepted view of a fuchsia. Their growth habits include a tree form, a crawling form and a shrubby type, yet all are true fuchsias! Species are becoming increasingly popular as growers examine the sources of today's hybrids and recreate some of the early crosses that form the foundation of modern fuchsias.

Above: Fuchsia denticulata, *with its beautiful, vivid flowers, was originally found in Peru and Bolivia and is easy to grow. The blue-tinged foliage has a shrubby growth habit, throwing up suckers around the main plant.*

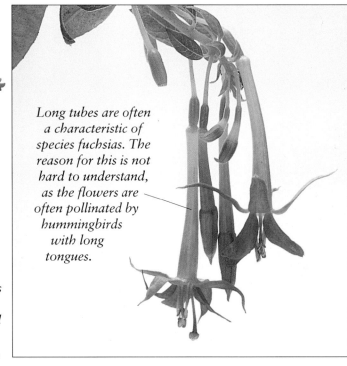

Long tubes are often a characteristic of species fuchsias. The reason for this is not hard to understand, as the flowers are often pollinated by hummingbirds with long tongues.

Above: The delicately scented flowers of F. arborescens *resemble lilac and grow in a terminal cluster. In Central Mexico it grows to 6m(20ft) tall, covered with flowers and fruits.*

Left: This is another species fuchsia from Peru - F. boliviana *var.* luxurians 'Alba'. *When it is in full flower, it can bear up to 40 flowers on one stem. This exceptionally strong grower can reach 4-5ft(1.2-1.5m) in one season.*

Below: Fuchsia fulgens rubra grandiflora *has incredible flowers up to 4in(10cm) long. It is a native of Mexico and will make a fine large plant that will thrive in a spacious container. There are a number of forms, but this is the easiest to grow.*

Left: Fuchsia procumbens *is one of the few species found in New Zealand. It has a prostrate habit. Pink fruits follow the flowers. This fairly hardy species is a delightful subject for a rockery.*

The triphylla hybrids

The triphyllas are a most distinct group of fuchsias that resulted from early crosses with *F. triphylla*, and as a group, they are vital to any fuchsia grower's collection of plants. The triphylla types are distinguished by a number of features: their distinctive, richly colored foliage varies from almost a metallic bronze to a rich purple sheen and their flowers have a long thin tube and range through pink, orange and red. They are able to succeed in a much more sunny and hot environment than most other fuchsias, and this allows them to occupy a wider range of positions in the garden. They thrive in large pots and tubs and will flower without a break for many months, given regular feeding and watering. Triphyllas are easy to grow. They do best in a stress-free environment that is neither too wet nor too dry; either extreme will cause them to lose their lower leaves. During the winter, keep the temperature above 45°F(7°C). Cut them back as you would any other fuchsia and they will soon grow again. Do not treat them as hardies unless you live in a warm area.

Triphylla hybrids

Bornemann's Beste, Coralle, Elfriede Ott, Gartenmeister Bonstedt, Leverhulme, Mary, Schonbrunner Schuljubilaum, Thalia, Traudchen Bonstedt, Trumpeter.

Below: A large container of Thalia - possibly the best known of the triphyllas. It is easy to grow and will thrive in a large container.

Triphyllas are characterized by dark to bronze-colored foliage and terminal flowers borne in clusters.

Left: Mary, one of the darkest triphyllas, has dark foliage and rich scarlet flowers. It was first introduced in Germany in 1894, one of many hybrids produced by Herr Bonstedt, who specialized in triphylla crosses.

Above: Sophie's Surprise, the first variegated triphylla, has typical flowers but most unusual foliage. From this it is hoped to produce many more variegated triphyllas.

Right: Elfriede Ott is another more modern triphylla. It is unusual in having semi-double rather than single flowers, i.e. five to eight petals, not four. Its lax growth makes it suitable for hanging baskets.

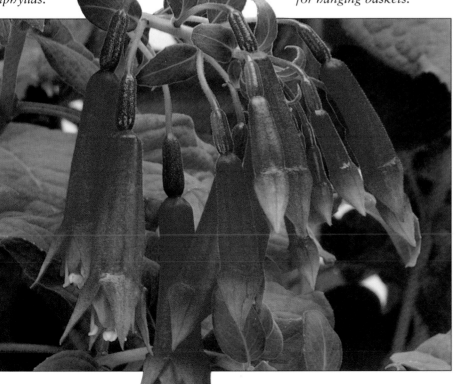

Left: Bornemann's Beste is another strong grower with terminal flowers. It will flower continuously for more than three months if you feed and water it conscientiously.

Right: Traudchen Bonstedt shows clearly that the flowers are borne on the end of the branches. This relates them to the original species, F. triphylla. Even though many crosses and hybrids have been produced, they still retain the characteristics of the parent species.

Encliandras

Encliandras are another group of fuchsias that deserve a closer examination. They have the most delightful small flowers, perfect in every detail, but never more than about 0.5in(1.25cm) long and frequently considerably less! As a group, they originated in Mexico and Central America and have a shrubby type of growth. Planted in the garden, they will make an enchanting variant in a hardy bed or in a summer bedding arrangement. A number of them are hardy and will survive the winter planted out in the ground in cool temperate climates. However, their greatest attraction must be their flowers. Look at them closely and you will find little jewels shining among the foliage. Their color range is not vast, but the assorted shades of white, pink, red and orange are more than sufficient. Much hybridization has been carried out to try and widen the range. In the wild, many natural crosses have occurred, as the species of this plant may occur with female-only or male-only flowers and some will be complete flowers, with both areas functioning normally. The plants with female-only flowers rely on hummingbirds and insects to pollinate them. These pollinators will have come from another encliandra and so there is a greater chance of cross-pollination than with mixed-sex flowers on the same plant. This is why there are so many crosses in existence today.

A selection of fine encliandras

F. x bacillaris, *Cinnabarina*,
F. hemsleyana, F. hidalgensis,
F. ravenii, F. thymifolia,
*Lottie Hobby, Miniature Jewels,
Neapolitan, White Clove.*

Below: A closer view of the exquisite flowers of Neapolitan. These flowers are red, which means they are fully mature. The buds are white and as they develop the flowers gradually become pink and finally turn red.

Above: *Cinnabarina is an example of the perfection of these tiny flowers. In this case, they are less than 0.4in(1cm) long and, as in most encliandras, the flowers arise from the axils of the leaves along much of the stem.*

Right: *Encliandras in the wild and in pots have a characteristic shrubby growth and will be covered in a multitude of small flowers for many months. Many of them will prove hardy when planted in the garden.*

Neapolitan is one of the more unusual fuchsias. At any one time it will have delicately perfumed, red, pink and white flowers.

F. *x* bacillaris *is closely related to the original encliandras from Mexico and Central America. It is thought that all the other encliandras have been crossed from this type.*

Above: F. *x* bacillaris *flowers are a lovely bright rose and, like all the encliandras, have a disproportionately large stigma compared with most fuchsias.*

Above and left: *Microchip is an interesting new hybrid with a wiry growth habit and masses of small, bicolored flowers that are unusual among the encliandras. It also has much paler leaves than most encliandras. The only true white-flowered encliandra is* F. hidalgensis, *which is a particularly rampant grower.*

Hardy fuchsias

Many fuchsias are described as hardy, which means that they survive the winter planted out in the garden in cool temperate climates and come up again the following year. Cultivars are only considered to be hardy when they are planted in the ground and allowed to form a natural shrubby shape; a basket or a standard will not be hardy, even though the cultivar may be considered as such. To a great extent, your climate will dictate what is hardy. In areas that are subject to regular frosts, and where fuchsias in the garden are regularly reduced in height, you will be more limited in your choice of plants. For example, you may buy a standard fuchsia described as hardy, only to find that in cooler climates it will be knocked back and only grows as a bush the following year. If you are uncertain as to what will be hardy in your area, make enquiries at a specialist fuchsia nursery, garden center or local fuchsia society. And remember, it is always worth experimenting, as you will often be surprised by what will survive. Generally speaking, neither triphyllas nor large exotic doubles will survive if planted out in cooler climates with a number of frosts during the winter. Put these subjects in a frost-free environment for the winter to be safe.

Right: Mr. A. Huggett is one of the first fuchsias to flower and gives a fine display for many months. Most fuchsias have an identifiable hybridist, but this one was found by chance.

Left: Reading Show, first introduced in the 1970s, has medium to large flowers of a good rich color.

Hardy fuchsias

Dwarf: *Alice Hoffmann, David, Lady Thumb, Son of Thumb, Tom Thumb.*
Medium: *Dorothy, Garden News, Margaret Brown, Rufus, Tennessee Waltz.*
Tall: *Edith, Margaret, Mrs. Popple, Rufus, Whiteknights Pearl.*

Right: *There are many hardies with interesting foliage and Sharpitor is one of the best. It has a delightful combination of foliage and flowers. This was a chance find in the garden of a National Trust property in western England.*

Strong colors in hardy fuchsias flowers are ideal to brighten up a dull area of the garden.

Above: *Beverley, a lovely vigorous grower, is another modern hardy. It is a single, with large, bell-shaped flowers. Modern hardy fuchsias are normally test-grown in a cold area to determine their hardiness.*

21

Standard fuchsias

Fuchsias look wonderful grown as standards; elevated on a stem you can enjoy the beauty of their flowers and their elegance so much more. It is an ideal way of looking at fuchsias with upright flowers. The basic requirements for growing a standard are a good, straight stem, a well-formed head and a good covering of flowers. Use a trailing cultivar and you will have a lovely weeping standard. You can grow any variety as a standard, although some will require much more effort than others. Never grow a standard that will be taller than you can easily manage to keep through the winter. Be wary of proportion; a large flower on a short standard will look odd, as will a very small head on a very long stem. Small flowers on small standards look much better. Always bring standard fuchsias into a frost-free environment for the winter months in cool temperate climates, otherwise you could end up with a bush rather than a standard. The stem can easily be caught by a cold spell, as there is no protection for it. All it needs to survive the winter is a warmer environment.

Right: *A young, first-year standard of Reg Gubler, a fuchsia named after the author's late father. Being a first-season plant, the head is still small. In the following year it will be larger and bushier.*

Right: *Royal Velvet, a fine old cultivar. The weight of the large double flowers gives it an almost weeping shape. This type of growth may need extra support, such as tying the branches to a central cane.*

The kink in the stem is unfortunate, but not so serious that it will cause a weakness in the plant's support.

Best cultivars

Singles: *Alison Patricia, Celia Smedley, Chang, Checkerboard, Estelle Marie, Jenny Sorenson, Joy Patmore, Olive Smith, Reg Gubler, Taddle.*
Doubles: *Annabel, Brookwood Bell, Cotton Candy, Dancing Flame, Happy Wedding Day, King's Ransom, Natasha Sinton, Paula Jane, Royal Velvet, Swingtime.*

Above: *Silver Dollar is a lovely white cultivar. Including plants at the base of a standard can considerably improve its appearance. The plants also help to keep the fuchsia cool by shading its roots.*

Right: *Shelford has been one of the most popular cultivars during the last few years. It flowers continuously for many months and is easy to shape. The shadier the planting position, the whiter the flowers will be. In a sunny spot, the flowers are almost pink.*

Fuchsias for hanging baskets

A hanging basket of fuchsias is wonderfully elegant and when it is in full flower it should be a cascade of blooms. Whether single- or double-flowered or grown for their foliage, they all produce equally phenomenal results. To achieve the ideal effect, which is for the growth to obscure the basket and have flowers cascading all over it, always use the same cultivar in one basket. If you use different fuchsias within the same basket, it will look messy and unbalanced. There are many basket cultivars available, all of which should have a lax style of growth. It is an ideal way to show off the very large doubles; the weight of their exotic blooms makes them ideal candidates for a basket hanging at a level where they can be seen. Single-flowered fuchsias, with their multitude of flowers, will also show themselves well in a basket. With these, you are more likely to achieve the ideal basket, with blooms right from the top to the bottom as a glorious waterfall of color. Remember that fuchsias look equally good in half baskets, whether they hang on the house or against a garden wall or fence. However, beware of placing any basket of fuchsias in a very sunny spot, as the warmth from a building can literally cook them. Place them on the shadier side and you will achieve far better results.

Right: An ideal basket cultivar is one that drips with flowers from top to bottom and where the depth of foliage is such that you cannot see the basket at all. Wilson's Pearls is an excellent example, with lovely, rich, semi-double blooms.

Careful attention throughout the summer will keep a basket like this in continuous flower for many months.

Left: Red Spider makes a basket, festooned with flowers for many months. Do not be deterred by the fuchsia's name, which may remind you of red spider mite; it produces an enchanting basket with ease.

Best cultivars

Singles: Abigail, Aunty Jinks, Daisy Bell, Hermiena, Jack Shahan, Marinka, President Margaret Slater, Red Spider, Waveney Gem.
Doubles: Applause, Blush O' Dawn, Dancing Flame, Devonshire Dumpling, Frau Hilde Rademacher, Malibu Mist, Pink Galore, Pink Marshmallow, Seventh Heaven, Swingtime, Wilson's Pearls.

Above *Waveney Gem is one of the most adaptable fuchsias. Not only does it grow into a superb basket, it can just as easily be grown into a standard, a large pot plant, a fan or a pillar.*

Right: *Abigail is a beautiful new Dutch cultivar, with exceptionally shaped flowers that open out into a saucer shape. To see it at its best, view it from below rather than from above.*

25

Bush fuchsias

The most common and the easiest way of growing fuchsias is as a bush, as this is most like the form in which they would naturally grow. Plant a fuchsia in the garden and nature will produce a neat, bushy plant. However, a plant in a pot does need some control and this is done by regular stopping, i.e. pinching out the growing tips to shape the plant. Ideally, a bush should be reasonably symmetrical and have an even development of growth and flowers. Pinching out gives the plant a good structure that will form the basis of its subsequent development. A good shape and structure will result in a plant that is full of flower and pleasing to the eye. If you feed and water it regularly it will give you many months of flowering pleasure. Once you have mastered the technique of shaping a bush fuchsia, you can try other forms. Basket and standard fuchsias, windowboxes, tubs and bedding are all shaped in the same way.

Left: Tom Thumb is well known as a dwarf hardy, but is bush-shaped. If a bush is defined as a plant with many branches but no trailing growth, then Tom Thumb is a fine specimen.

Right: Jenny Sorenson is a delightful single-flowered fuchsia with delicate banding around the edges of the petals. It branches out well and has a stiff growth habit.

Single cultivars of bush fuchsias

Bill Gilbert
Border Queen
Carla Johnson
Jenny Sorenson
Love's Reward
Mission Bells
Nellie Nuttall
Pink Fantasia
Shelford
Taddle

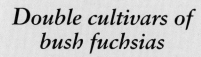

Right: *Voodoo is a double with large flowers borne on strong stems. Unlike many doubles that need to grow in a position where they can trail, Voodoo will develop into a fine bush, producing a good number of splendid flowers.*

Below: *Taddle is a single-flowered fuchsia that grows into an excellent bushy shape if regularly stopped. This short-jointed, compact plant will flower continuously for a period of many months.*

Double cultivars of bush fuchsias

Annabel
Brookwood Bell
Eusabia
Heidi Ann
Marcus Graham
Paula Jane
Ridestar
Spion Kop
Topper
Voodoo

Above: *Annabel has almost white, large, showy flowers, but they are not too heavy to weigh down the stems of this sturdy bush. Pinch out regularly.*

Fuchsias for foliage

When considering which fuchsias to plant, it is tempting to choose varieties purely for their flowers, but why not consider the beauty of their foliage? There are many cultivars of fuchsias with ornamental and variegated foliage just waiting to be discovered. Foliage fuchsias will add color to your garden long before they are in flower. They are all mutations, or sports, from a cultivar with ordinary green foliage. This means that any flowers they produce will tend to be smaller than the parent cultivar, as they are in reality a weakened form of the original. Even so, the flowers and foliage together will really enhance the plant. The old foliage fuchsias tend to have red-purple flowers, whereas the more modern ones can bear any type of flowers, from small singles to large, exotic doubles. When they flower, the effect can be stunning. Some of the earliest variegated fuchsias originated in the 1800s and are still very popular. Today, with the great influx of new cultivars in the last 20 years or so, the number of ornamental-leaved fuchsias has increased dramatically. As well as the variegated-leaved fuchsias, why not add even more to your color range with the triphyllas and their dark velvety leaves? Or try *F. magellanica aurea*, with its lovely golden foliage that will make a bright splash of color. The depth of the foliage color will be enhanced by growing the plants in the shade and, as usual, by good and regular feeding. The variety Tom West will become much pinker in the shade, while Genii is much more yellow. Foliage fuchsias are ideal for growing in any sort of container, as they will add color and variety to your planting schemes all year round.

Golden Marinka, a variegated form of the lovely old basket cultivar, has deep red single flowers. It will also make a graceful basket.

Fuchsias for foliage

Every collection of fuchsias should include some with attractive foliage. Good choices include the following varieties: Cloth of Gold, F. magellanica aurea, Genii, Golden Marinka, Ornamental Pearl, Pop Whitlock, Sharpitor, Sunray, Tom West, Tricolor, Tropic Sunset.

F. magellanica aurea is a golden form of the strong-growing hardy. It is just as hardy and will give its best golden color when planted in a shady spot.

The variety Tom West was first discovered in 1853. This absolutely delightful fuchsia, with its shiny foliage, is ideal in hanging baskets and troughs.

Popsie Girl is a new variegated form of Orange Drops, with a matt finish to its foliage. The flowers are bright orange. Examine your plants closely for any change in the foliage; Popsie Girl was found in this way.

Autumnale, or Burning Bush, has glowing gold and red leaves and grows strongly. It will fill a hanging basket or flourish in a pot.

Tricolor, a strong-growing garden hardy with three distinct tones to its leaves, will brighten any flowerbed. This old cultivar is a variegated form of F. magellanica.

29

Choosing a good plant

If you are just beginning to investigate the world of fuchsias, you will want to find the best places to start or increase your collection. There are a number of options, but a specialist nursery has to be the best. The people there will be able to advise you about choosing the most appropriate varieties and how to grow them. Garden centers can also be good sources of plants, but the range of varieties may be limited. Finally, it is always worth investigating local plant sales, where gems are often to be found, even if they are not correctly named. There are various points to look for before you actually buy a plant. Firstly, examine it closely to make sure it is healthy. If the plant dies within a few days of purchase, then the chances are that it was not healthy at the nursery. It is very much a case of buyer beware; look under the leaves and at the stem for any possible damage or pests. For example, if the base of the stem is a shiny brown, it could be that the plant has had problems with botrytis at some stage, and the stem will always have a weakness, so look for another plant.

Above: Look underneath leaves as well as on top for signs of pests or disease. Plants with yellow leaves may be potbound or underfed.

BAD PLANT
This plant has been pinched out too far up the stem and is a poor shape. It would be difficult to do anything constructive with such a plant.

BAD PLANT
There are signs of rust on the leaves. This problem spreads quickly and could affect your other plants.

BAD PLANT
This plant has been stopped once and is becoming straggly. It may have had too little light and has also become potbound.

BAD PLANT
A generally neglected plant with signs of rust. A hard pinch back will improve the shape and a fungicide will prevent the return of rust.

GOOD PLANT
A fine bushy, healthy plant immediately catches the eye. A nursery selling plants like this is worth a visit.

Taking a chance

The only time you might wish to consider buying a plant that looks in dubious condition is if it is a variety that you are desperate to acquire. In that case, take a cutting for insurance as soon as you return home. If the original plant has a poor shape, cut it hard back. The new growth should be healthy and you can improve its shape by regular stopping. If you suspect that the plant is incorrectly named, wait until it is in flower before seeking expert advice.

Right: *If you buy fuchsias in flower, make sure they look healthy. Ideally, they should just be coming into flower so that if they drop any buds there will be plenty more to follow. This specimen of* F. arborescens *has healthy foliage, has been well-shaped and promises more attractive blooms.*

BAD PLANT
A thoroughly neglected plant without even a hint of shape. It might be worth taking cuttings from it.

BAD PLANT
This larger plant has been stopped once and then allowed to grow on. The branch on the left has lost leaves, perhaps due to damping off earlier on.

GOOD PLANT
This fine bushy fuchsia will make a good plant. It has been regularly fed, turned and pinched out.

Choosing the right soil

There are many planting mixtures available; some are peat-based, others contain loam, coco-fiber or even bark. Fuchsias are very amenable and will grow in any medium, as long as it is free-draining and contains nothing harmful. Look in your garden center or nursery at the many products available. A premixed product from one of these outlets is an excellent choice, as it contains everything your plant could need, including feed. Check that stocks are fresh; old mixtures tend to smell damp, musty and stale. Making up your own potting mixture is probably the best option; not only will it always be fresh, but you will know exactly what it contains. The main ingredient of a planting mixture should be a growing medium, such as peat, to retain moisture and nutrients and allow for drainage. You will also need a product to open up the growing medium and increase drainage; this could be some kind of grit, coarse sand or perlite. Finally you should add a slow-release fertilizer, which is available as a prepackaged product and contains an excellent balance of ingredients, including trace elements, to promote healthy growing plants. It may be sold as a general fertilizer that you add to the potting mix; if in doubt, ask for advice, as there are many products to choose from. Keep the potting mixture dry to prolong its useful life and only make up as much as you need, when you need it.

Right: When grown in a suitable rooting medium, a plant will soon develop a collection of small, fibrous roots and a central tap root. Healthy roots are white with very tiny hairlike protrusions that increase their surface area and their ability to take up food and water.

Mixtures that contain no peat are increasingly popular. This one is made up of bark and other woody material. Some contain coco fiber, cocoa shells and even paper waste.

In many areas, peat-based mixes are still easy to obtain and are an excellent basis for a planting mixture. They are soft and extremely fibrous and they retain water and nutrients well. Fuchsias thrive in this medium.

Add one 6in (15cm) potful of grit to every bucket of peat and other fibrous planting mixtures. It improves drainage and adds weight.

Perlite is a very good alternative to grit for improving drainage in peat-based mixtures. Being light, it is ideal for hanging baskets.

Vermiculite is able to retain water, as well as open out the texture of a planting mixture to improve drainage. Add it on its own or together with perlite or grit.

Slow-release fertilizers added to the basic mixture are an excellent way of ensuring that your plants receive vital nutrients. Use a liquid feed as well.

Traditionally, fuchsias were grown in a loam-based medium. This is a much heavier product than the others and ideal for plants in an exposed situation and buffeted by winds

Terracotta pots are slightly porous and ideal for all fuchsias, particularly the triphyllas that prefer drier conditions.

Fuchsias do well in plastic pots, but shade the roots. If necessary, put the pot inside a larger one; the air trapped between the two pots keeps the roots cool.

Water-retaining gel

In warm and quick-drying conditions, these gels help to retain moisture in peat and other non-loam planting mixtures.

1 Add the granules to water and leave them until they expand into a glutinous mass.

2 Add the required amount of gel to the soil. Too much gel could cause waterlogging.

3 Mix the gel in thoroughly, carefully following the manufacturer's instructions.

Planting hardy fuchsias

1 *Remove the fuchsia from its pot. Put some peat in the base of the planting hole. Plant the fuchsia 1in (2.5cm) deeper than it was in the pot.*

Fuchsias are ideal for borders in the garden and, once established, will delight you for many years. Once you have chosen the planting area, add some humus and some slow-release fertilizer to the soil to encourage the plants to grow well in their first season. Before planting out hardy fuchsias, make sure you know how tall they are likely to grow, and base your planting plan around those heights. Nurseries and garden centers will be happy to advise you if you are in doubt. Planting is a simple operation, but be sure to follow it up correctly. Water your plants regularly during the first season to help them establish a good root run, which is vital for their survival. After the first year, watering will depend on the climate in your area. Never water a plant when it is in full sun, as water splashed onto the leaves could cause them to scorch and burn. One way to reduce the need for watering is to apply a mulch, such as chipped bark, around the plants and this also helps to keep down weeds that might compete with the fuchsias. An occasional feed during the first season helps to build up a plant so that it can survive its first winter. Before the onset of the colder weather, give your hardy fuchsias a little extra protection, either by earthing them up or by placing straw or some other suitable insulating medium around the plant. This will keep them warm, which is particularly important during their first winter in cool temperate climates.

2 *Support the fuchsia gently while you fill in the hole with the excavated soil. Take care not to damage the rootball. Water the fuchsia before planting it out.*

3 *Press the soil down gently with your hands. Make the area around the actual plant slightly higher, as this will help the plant survive the winter.*

White Pixie, a red-and-white single.

Brutus, an old single cultivar.

Nicola Jane has large double flowers.

Left: *Position your plants carefully to allow them plenty of room to grow. Plant the shorter ones at the front and the taller varieties at the back. There are fuchsias to suit every aspect.*

Alice Hoffmann has rose-and-white semi-double flowers.

5 Add a little mulch around the plant. This retains the moisture in the soil and reduce the chances of any weeds growing around the plant and competing for nutrients.

Pink Goon, a lovely double. (The same variety is planted at bottom right.)

F. magellanica aurea

Son of Thumb, a shorter grower.

4 Water the plant without getting moisture on the leaves. Keep the newly planted fuchsias well watered during their first year while they establish a good root run.

Growing standard fuchsias

People often view the prospect of growing a standard with great trepidation, whereas it is no harder or easier than any other form of growth. Virtually any variety of fuchsia can be grown as a standard: trailing varieties make a wonderful weeping head, while bush varieties result in a more traditional, but equally delightful, type of standard. A standard is really a well-stopped plant on top of a stem that is as straight as possible. Any flaw or blemish in the stem can reduce its strength, which can cause problems if the plant is in an exposed position. It is harder to produce a very straight stem on a trailing fuchsia, as the growth is naturally more supple, but regularly tying the stem to a cane can help. When selecting a cutting to grow as a standard, try to choose one with three leaves at each node, as this will produce a superior standard to a cutting with just two leaves. Tie it onto a cane as soon as possible, so that the plant gets off to a good start. Pot the whip, as it is now called, into a larger pot as often as necessary; a potbound plant will start to flower and not grow, thus defeating the object of the exercise. The final height of a standard is up to you; generally speaking, the taller ones have more problems and can look out of proportion until they are several years old and have achieved a good size head. Bear in mind that tall standards can be difficult to store during the winter. All standards must be kept frost-free, otherwise the stem can die and you will be left with a bush the following year.

5 *Now the remaining five side shoots will start to grow. Leave them to produce two or three pairs of leaves. Pinch out the growing tips.*

1 *Choose a cutting with a straight stem and, if possible, three leaves at each node. Carefully insert a stick into the potting mixture and prepare to tie the cutting to it.*

2 *Gently bring the cutting in close to the stick. Use a soft tie to avoid damaging the tender growth. Keep the stem as straight as possible; it is particularly supple at this stage.*

3 *Carefully break off any side shoots that appear in the axil of the leaves by gently bending them to one side. Retain the top five sets of leaves, as they will form the head.*

4 *When the young standard has reached the required height, remove the growing tip with a pair of sharp scissors. Do not remove the large leaves on the main stem yet.*

8 Now that it has a well-formed head, the young standard is ready to flower. Do not allow it to flower until it has reached this stage, as the weight of the flowers, particularly large doubles, can break the stem.

A good, straight stem will make for a stronger standard.

A standard emerges

As standards develop, they need attention every few days. A plant can produce a remarkable amount of new growth in a short time. A regular care program produces better plants.

A young standard, ready for its first stop on the side shoots.

A tall plant, with as many leaves as possible left on the stem.

A cutting just after the first tie-in.

6 As the bushy head starts to form, remove the large leaves on the main stem, which has become quite woody at this stage. Small side shoots often appear on the main stem away from the head.

7 Gently break off any side shoots, taking care not to damage the main stem. It is important to remove these side shoots, otherwise they are wasting the plant's energy.

Planting a basket of fuchsias

Most people would agree that the ideal way to view fuchsia flowers is when they are growing in a hanging basket. You can really see them in all their glory from below or, best of all, at eye level. For the best effect, restrict each basket to only one variety of fuchsia; if you mix them, they begin to look messy, as different varieties will grow at different rates and flower at different times. Eventually, the whole effect can become untidy. The number of plants that you put in a basket depends very much on the size of the basket. Always plant one in the middle, otherwise you can end up with a hole as the plants begin to grow downwards. The plants still need regular stopping after every two or three pairs of leaves to achieve a longer, but not straggly, look. Choose your varieties with care and you cannot fail to be delighted with the effect that your basket will give, whether it drips with enormous double flowers, or cascades from top to bottom with smaller single flowers. To ensure that the basket flowers continuously for many months, remove dead flowerheads and seedpods and feed it regularly. A slow-release fertilizer is useful if the basket is difficult to reach for liquid feeding. Remember to turn the basket regularly to make sure that it retains a balanced shape as it grows.

3 Finally add the central plant. This one will prevent a hole appearing at the top of the basket once the plants have started to grow.

1 Choose a peat-based potting medium as this will make the basket lighter when you hang it up. Select a suitable cultivar for your basket and always use an odd number of plants for best effect.

2 Place a little potting mixture in the base of the basket and then position all the plants except one around the edge of the container in a symmetrical pattern.

4 *Fill in the gaps between the plants with soil, but do not push it down - let it find its own level. Add a slow-release fertilizer to make feeding easier.*

7 *The basket of Ballet Girl is coming into flower 10-12 weeks after the final stop. The blooms are hanging over the sides of the basket and will soon cascade in all directions.*

5 *Fix the chains onto the basket, so that the plants can grow around them without being damaged. Tie the chains securely to a cane to keep them well above the growing plants.*

6 *After six weeks the plants have filled the basket and are beginning to trail over the sides. This basket had its final stop three weeks before it reached this stage.*

Feeding and watering

Above: Yellowing leaves are often linked to a lack of feed. Give plants a general balanced feed with a variety of trace elements.

Maintaining a regular feeding program during the year will make all the difference to general growth and flowering. The feeding routine can be quite simple: every 10-15 days in spring, use a high-nitrogen feed with a proportion of 25:15:15 (nitrogen: phosphate: potash) to encourage healthy growth. Never apply a stronger mix than is recommended by the manufacturer; too much high-nitrogen feed can result in soft, lush growth. Change the feed in late spring to encourage the plant to begin producing flowers. Aim for a balanced feed with a proportion of 20:20:20 and stay with this until the end of the flowering season. Again, never overdo the quantity of feed in any one week. If you prefer, feed the plants with the fertilizer diluted to quarter strength four times a week or twice a week at half strength, for example. Simply choose a plan that suits you best. If flowering is a little slow, it can be worth using a feed, such as a tomato fertilizer, that has a higher potash content, i.e. 15:15:30. However, good daylight and regular, balanced feeding should be sufficient to produce a plant that is full of flower. Water fuchsias individually, as some will require vast quantities and others seem to need very little.

Left: A lack of water results in soft, limp leaves and wilted growing tips. Water the plant gently and empty the saucer after 30 minutes, rather than leaving the plant standing in water.

Above: Gentle spraying helps to create a humid atmosphere and certain feeds are also applied in this way. Look for foliar feeds; the plants take up the nutrients through their leaves.

Below: *Liquid feeds are very popular. Follow the instructions on the bottle and do not be tempted to feed more than a full dose. It is much better to feed plants at half strength twice a week than to overdo it.*

Watering fuchsias

Overwatering causes as many problems as underwatering. A very wet fuchsia will look as limp as a dry one - only the wetness of the soil will give you a clue. If you think a plant is too wet, place it in a cool, shady spot. If the soil is saturated, take the plant out of its pot and place it on dry blotting paper to draw water out of the soil and prevent the plant from drowning.

Well-fed, healthy foliage has a good, strong color.

Below: *Keep powdered feeds in a closed container, as they take up moisture from the air. Keep the feeding program simple - your plants will be the better for it.*

Right: *Slow-release feeds have become very popular. Feed is released from a capsule over a period of months. This is ideal for hanging baskets and tubs that are not always easy to feed.*

Put the feeding pellet in the pot near the rootball. A large container, such as a hanging basket, may need two or three pellets.

41

Summer care

Once fuchsias are in flower there is a tendency to neglect them, but this is just the time that they need your closest attention. Feeding is particularly important during the summer, as providing a balanced feed once a week will promote continual flowering over a long period. Carry out regular maintenance once a week - more frequently if possible - to ensure that the plants continue to flower for many months. Take off all the seedpods, dead flowers and yellow leaves and check that there are no problems with the plants. Vigilance and care during the summer months will ensure that you spot any problems caused by pests or disease at an early stage.

Watering is vital at this time of year, otherwise fuchsias will lose their leaves and flowers. However, do not overwater; check the dampness of the soil by feeling the surface with your fingers. It should only ever be damp to the touch, never soggy. In the hottest weather, fuchsias may need watering several times each day. If it remains hot for prolonged periods, it could be worth installing a drip-feed watering system that supplies a trickle of water to each plant through a network of small-bore tubes laid on the ground. Plenty of shade should reduce the need for watering, as does a humid atmosphere, so spray water on the ground around the plants as you water them.

Right: When fuchsias are planted in a garden border, nature will take care of most of the tidying and maintenance tasks for you. Wind and rain will knock the old flowers off the stems.

Bees' footprints

Insects visiting flowers to collect nectar may gradually damage the petals. Examine the flowers closely and you may find small, dark patches on them. If you wish to show your plants, move them to a spot where the insects cannot reach them.

Water border fuchsias regularly, as they have not had time to establish a full root run. Feed them several times during the growing season.

Left: Tubs require more care in summer. Remove dead blooms, seedpods and yellow leaves every week to ensure the longest possible flowering period. Feed the tubs once a week in summer. When feeding and watering, take care not to get liquid on the leaves, as this can cause scorching.

Above: Fuchsia flowers that are almost dead come away easily if you pull them gently. In some cultivars, dead flowers fall off readily, while in others they hang on until removed.

Left: Remove yellowing, dead or damaged leaves to improve the general appearance of the plant. At the same time, examine the plant for any signs of pests and diseases.

Right: Remove any seedpods that remain after the flowers have gone. If they have been fertilized and are allowed to develop into mature fruits, the plant is less likely to continue flowering.

Pruning basket fuchsias

As fuchsias approach the end of their growing and flowering season, the foliage starts to turn yellow and there are few new flowers. This is the time to give them a rest and to create a plant with a good shape and structure for the following year. Never be afraid to cut back a plant. Baskets and standards will do better the following year if you have cut them back carefully. Shape the plants as you cut until you are happy with the woody structure. Strip away any leaves that remain on the stems and remove any debris on the surface of the soil. Remove the plant from its pot and check that no vine weevils are present. Examine the roots for evidence of chewing, patches of missing roots and holes in the soil. If you find any of these, excavate gently with your fingers until you find the vine weevil larvae and remove them. Label your plants while you can still remember what they are or just make a note of the colors. If you leave the fuchsia in the same potting mixture during the winter, and do not disturb the rootball, either repot it into the same basket in spring or leave it as second-year basket in the same soil. Add a little fresh potting mixture on the surface and feed plants regularly.

2 Aim to create a sound structure for the following year, but retain a sufficient amount of growth so that any possible die back in the winter will not endanger the final shape next season.

1 At the end of the growing and flowering season, hanging baskets lose all inclination to flower. Cutting them back at this time will force them into a period of rest. Any subsequent new growth will be fresh and full of vigor.

4 You can see the finished effect and even make out the shape if it is unbalanced. Label all the plants, as it is easy to forget which is which when they resume growing in the spring.

3 Leave the main branches long enough so that they are already beginning to hang over the edge, ready for next season. Use sharp secateurs so that you cut rather than tear the stems.

When dealing with round-bottomed baskets, sit them in a bucket to stop them rolling around.

6 Once you have finished, keep the basket in a frost-free environment. In areas where the temperature never drops very low, cutting back ensures that the plant gets a period of rest.

Shortly after cutting, the stems may bleed a little with a clear liquid. This will soon stop and does not harm the plant.

5 Remove any leaves that remain on the stems. Clear away debris on the surface of the potting mixture to prevent the development of problems with mold, etc., during the winter months.

45

Cutting back a standard fuchsia

Patience, a medium-sized double with white-flushed, pale pink flowers.

1 *At the end of the season, fuchsias start to look tired, with yellow leaves and a clear lack of flower. This is an ideal time to give the plant a rest and prune it back.*

When a fuchsia still has a few buds and green leaves left on it, it is very tempting to hesitate about cutting it back. However, failure to cut back will mean that over a period of time, the standard will deteriorate. The head will grow larger and larger, but the flowers and growth will become more and more sparse. With good care and an annual cut back, standard fuchsias can survive for many years - a heavily pruned standard has been known to live as long as 20 years. As you cut back the standard, it is important to visualize the final shape that you are aiming to achieve. By careful cutting you should be able to create a woody structure that is at the very least as good as the current year's growth. With careful trimming, you should be able to create an even better woody shape that will pay dividends in the following growing season. Where possible, cut above a fork in the branches, so that as many branches as possible will develop.

If you prune in warm conditions, there is little delay between cutting back and the appearance of new growth. Within 10 to 14 days, small specks of new growth will start to sprout. The speed at which they develop will depend on the climate. In temperatures of over 50°F(10°C), fuchsias continue to grow and you will soon see a fine new head on a standard.

2 *When pruning a standard fuchsia, the aim is to produce a good, woody shape and structure for the new growth. Use a sharp pair of secateurs.*

3 *Work your way around the plant to create a well-balanced shape. Err on the cautious side; you can always cut off a little more later on.*

Try to cut above a division of branches to maintain a more substantial structure.

4 *As you prune more of the old growth away, the shape of the existing woody structure becomes clearer. This is a good opportunity to make a lop-sided plant more symmetrical.*

Do not cut off any more growth on this side.

5 *It is amazing how much you will cut away as you work, and how much of the plant you can see. Ideally, the head should be no more than a third to a half of its original size when you have finished.*

6 *Once you have finished cutting, remove all the leaves left on the wood, otherwise you may encounter problems with fungal infections, such as botrytis and damping-off.*

7 *Now there is a good framework for the new growth in the following season. There may be a little leakage of sap, but this will soon stop and does not harm the plant.*

Preparing your fuchsias for winter

How you prepare fuchsias for the winter depends very much on the severity of the weather in your area. If frosts are extremely rare, you need not take many precautions, but your plants will still need a dormant period, otherwise they become weak and will not give you the best results. Cut back your plants as described on pages 44-47. They will continue to grow unless a cold spell occurs and checks their development, in which case you may find it useful to keep some lightweight fleece to hand for added protection. Fleece is available from nurseries and garden centers.

Assuming you live in an area with a relatively harsh winter and a number of frosts, these are the precautions you should take. Once a plant has been prepared, move it to a heated greenhouse or an alternative frost-free environment. If it is kept below freezing point for more than a short time, a plant will die. At 32-39°F (0-4°C) the plant will survive and not grow. Above this temperature the plant will resume growing. Never let your plants dry out entirely during the winter; always check that the soil is damp.

2 The aim of cutting back is to produce a good structure of woody stems, half the size of the original plant. Now is the time to correct any defects that have appeared in the shape of the plant.

Use a good sharp pair of secateurs to enable you to make clean cuts into the wood rather than tearing it, which could encourage disease.

1 Plants that are coming to the end of their growing season start to look tired and have yellow leaves. They need to rest in winter; cutting back forces them to stop growing.

3 The more wood you remove, the clearer the structure that you are aiming for will become. Remove the largest branches first.

4 Once the largest branches have been cut back, you can look at the smaller branches that remain and then cut them to create the shape that you want.

5 When you have achieved the desired shape, remove any leaves left on the stems and clean away any debris on the surface of the pot to prevent botrytis and damping-off.

Above: A cardboard box filled with polystyrene foam chips is excellent insulation and allows the plant to survive the winter. Surround the plant to keep it as warm as possible.

The fleece allows the light through, so plants will grow underneath without being damaged.

Right: In an area that is almost frost-free, lightweight fleece will give the plants a little extra protection. A single layer will give protection to 29°F(-1.5°C), a double layer will give several more degrees of protection.

Starting up in spring

As soon as the air and soil temperatures begin to rise, your plants will be stimulated into the early signs of growth. If they have been kept in a dark environment, then any young growth will be white and straggly and will fall off soon after it has been brought out into the light. However, this will soon be replaced with healthy green shoots. Light and warmth are vital during this period of the year, and a good mixture of the two will encourage the plant to grow. Once new growth has been established, the time has come to have a close look at the plant. Firstly, has there been any dieback on the old wood? If so, lightly prune the plant again, but be sure to keep an idea of the ultimate shape of the plant firmly in mind. Spring is an excellent time to give fuchsias some fresh potting mixture to stimulate them into even more growth. If you can get the plant into a slightly smaller pot you will have an opportunity to pot it back up into a larger container later in the season. It is equally important that you begin to feed your plants at this stage; a high-nitrogen feed is the most suitable, as this will stimulate growth (see page 40-41).

2 Even in spring this plant has plenty of healthy roots. If there are few roots, check for vine weevil larvae. Old brown roots mean the plant needs fresh potting mixture.

1 Fuchsias kept at above 40°F(5°C) continue growing slowly through the winter. As the days lengthen, growth speeds up. Pot them into some fresh soil to give them a boost.

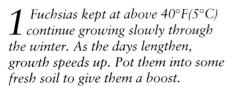

If potting down to a smaller pot is not feasible, use the same size pot as before.

Potting at this stage allows you to do further potting later in the year into larger pots.

Spring growth should appear from the base of the plant to the tips of the wood. If not, prune any dead wood.

3 *Remove both old roots and potting mixture with your hands. Do not worry if you damage the roots; they will soon grow back and be even stronger than they were before.*

4 *In order to fit the plant back into the same size pot or a slightly smaller one, you must be ruthless - but still leave a good mass of roots.*

5 *Place a little fresh potting mixture in the bottom of the pot. Position the plant in the pot and gently add fresh soil down the sides of the rootball. Do not press down the potting mixture.*

If a small standard does not grow from the top, trim off the top and grow it as a tall bush.

Right: *Spray the wood with tepid water daily to soften the bark and stimulate the plant into growth.*

Below: *If a standard is slow to grow, lay it down to make it easier for the sap to get to the top.*

Slow growers

If a fuchsia is not growing, gently scrape off a little bark with your fingernail. If it is green underneath, the plant is alive. If it is brown, scrape another area of bark, as the plant may have died back a little. If it is alive, it will need some extra nurturing to stimulate it into life.

This variety is Auntie Jinks. It has been successfully potted down into a smaller container.

6 *If the potting mixture sinks down too far, add some more to the top at a later stage. Water the plant and keep it in a stress-free environment for several days.*

Pruning hardy fuchsias

All hardy fuchsias require pruning at some stage, even if you live in an area with few or no frosts, otherwise they become tired. When you prune them depends very much on the local climate, because you must wait until there is no danger of frost that could damage the young growth. If in doubt, delay pruning, but then attack with enthusiasm! In areas where frosts occur, plants will die back to ground level. In this case you must wait until the new growth has reached 2-3in(5-7.5cm) high before you consider cutting off the dead wood. Where a plant is slightly damaged by the colder weather but new growth appears part of the way up the stems, adopt a more cautious pruning method. When you cut back these plants, think along the lines of cutting back a bush and aim to create a good shape for the new growth. Carry out a less enthusiastic but firm cut just above the new growth. However, if you do not want the plant to get too large, cut it back more firmly. Finally, in areas where frosts never or rarely occur and the plant has no reason to die back at all, it is still a good idea to prune hardy fuchsias once a year. Prune the whole plant in one go or, if you prefer a less drastic method, carefully prune one third of the branches and another third later in the year.

1 *It is important to remove the old wood that has been killed by the frosts. Use sharp secateurs to make clean cuts. Work carefully, so that there is no risk of damaging the new young shoots.*

Right: *If they have been caught by the frosts, hardy fuchsias will need pruning in the spring to remove the dead wood. In areas without frosts, fuchsias will still need partial pruning to ensure that they give a good show.*

Dorothy, a hardy with beautiful bell-shaped flowers. It makes a low-growing mound about 24in (60cm) high.

Cut low into the old wood so that it does not look unsightly.

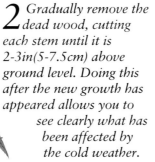

2 Gradually remove the dead wood, cutting each stem until it is 2-3in(5-7.5cm) above ground level. Doing this after the new growth has appeared allows you to see clearly what has been affected by the cold weather.

Hints on pruning

Never cut fuchsias back before the frosts have come; leaving the dead twigs on the plant will help to break up the frost and increases the plant's chances of survival during the coldest times. Earthing up the crown of the plant with soil or leaves also helps. Start cutting back only when the danger of hard frosts in your area is past and when the new growth is showing strongly. If the plants have not been killed right back, prune off only the dead wood, leaving the new growth intact to give the plant a greater size than before.

Right: Cardinal Farges is a bright, hardy fuchsia, originally from France. The flowers are semi-double and it has a compact growth habit. As with all fuchsias, pruning encourages it to produce healthy, new growth.

3 As you remove the last stem you can see that a new bushy plant is emerging with fine healthy growth. A little feed at this stage will give the plant a good start.

Young healthy growth on Pixie, which has light foliage and rose and light mauve flowers.

4 The fuchsia will return to its original size within weeks. This growing method means that you need not worry about turning or stopping - nature will create healthy, bushy plants.

Pests and diseases

Luckily, fuchsias do not suffer from too many pests and diseases, but it is worth familiarizing yourself with potential problems so that you will know what to look for. If a particular problem is not covered here, consult your local nursery; they will be happy to help you.

Whitefly is probably the most common problem wherever you grow fuchsias. Whiteflies are all female and can produce subsequent generations at incredible speed. Over the years, they have become resistant to most sprays, but *Encarsia formosa*, a minute predatory type of wasp is very efficient at eliminating them.

Aphids, particularly greenfly, are very much a seasonal problem. Left to multiply, they will gradually distort young growth on the plant. A number of chemical sprays are available and these will normally solve the problem. There is also a predator, *Aphidius matricariae*, but it is relatively large and mobile and, unless your plants are in an enclosed area, it is likely to fly away.

Red spider mites are more of a problem, as they are so small that most people do not see the pests, just the damage that they cause. The first signs of the mite are small white dots on the upper surface of the leaf, but again they are extremely difficult to see. Tackle the problem either with sprays or a natural predator, *Phytoseiulus persimilius*. Also remove and destroy any damaged leaves and isolate affected plants. Red spider mite tends to thrive in a dry environment, so promote humid conditions to discourage it.

Although adult vine weevils do not cause fuchsias too much damage, their larvae can create serious problems and you should check your plants regularly for signs of their presence.

Rust is a fungus problem that will never kill a plant, but can cause severe problems if left unchecked. The fungus is found within the system of the plant and will produce areas of fruiting bodies that produce vast quantities of spores on the underside of leaves. It thrives where plants are kept close together and the ventilation is poor, so improving these conditions will reduce the chance of rust. Remove and destroy any affected leaves and, if the problem is severe, do the same with the top layer of soil. Spraying regularly with a fungicide will control the problem.

Finally, if you are using any sprays, follow the manufacturer's instructions carefully and always take any necessary precautions. Never mix sprays or increase the dose. If in doubt, ask for advice.

Right: Using a magnifying lens you can see the small red spider mites on the underside of the leaf. They pierce the leaves and suck out the contents of the cells, causing discoloration.

Above: This plant is suffering from a particularly heavy infestation of red spider mite. Even the tops of the leaves are affected. They are turning brown and beginning to roll under.

Vine weevils

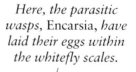

Vine weevil larvae (shown far right) will kill fuchsias if not controlled. The adults (shown right) are a minor problem, as they only chew the edges of leaves, leaving small horseshoe-shaped indentations. Being nocturnal, they are not easy to spot, but if you find one, tread on it! Vine weevils are all female and lay their eggs in the top of the soil during the summer months. These hatch and become larvae, with the sole aim of eating roots (not only fuchsia roots but those of many other types of plant as well). In the fall, you can often spot early signs of damage by taking the plant out of its pot and examining it for chewed and brown roots. Look in the area of damage and you will find the larvae; remove them and feed them to the birds! Alternatively, control them biologically, using a nematode worm called Heterorhabditis *sp.*

Above: Rust occurs in warm, poorly ventilated conditions. The spores on the leaves are easily dislodged and transferred from plant to plant.

Scales contain the immature whitefly.

A mature whitefly resembles a tiny moth, less than 0.04in(1mm) long.

Here, the parasitic wasps, Encarsia, have laid their eggs within the whitefly scales.

Below: A single aphid. These insects multiply rapidly. They suck out the contents of leaf cells, causing damage to immature leaves and growing tips.

Above: The only way to treat the plant is to cut it hard back and burn the plant material. Water the potting mixture with an appropriate chemical to destroy any remaining mites.

Above: All the stages of the whitefly problem on one leaf. The adults are attracted to yellow; you can buy sticky yellow 'traps' to combat infestation.

55

Taking a cutting

The best time to take fuchsia cuttings is when the growth is soft and green. Avoid taking cuttings during the summer months, when it tends to be harder and woody. Furthermore, once a plant is determined to flower, it is less likely to root. Having prepared the ideal environment for your cuttings to root and grow (see pages 58-59), the next crucial step is to choose the right plant material. If you start with a poor cutting, it is much harder to grow a good plant. Choose young and tender growth, as old, woody material takes much longer to root. Remember that the growth hormones that control the plant's ability to root and grow are found in the very tip of the stem, so long cuttings will take more time to root. A good cutting should look healthy and have two leaves of the same size around the stem. Equal size is important to produce a fine balanced plant. Some plants occasionally produce stems with three leaves rather than two, and these make excellent cuttings and a far superior plant. However, such cuttings are the exception rather than the rule.

Finally, have everything prepared before you start to take cuttings, as a delay between cutting and placing the pieces in the rooting medium of your choice can be fatal to the delicate young plants.

Below: This plant is covered with young growing tips, which make excellent cuttings. In taking cuttings, you are performing two tasks at once: increasing your stocks and improving the shape of the plant by stopping it.

This soft, healthy growth should root successfully and provide you with a supply of fine new plants.

These stems are becoming woody. Cuttings taken from such a plant will take a long time to root.

Small flowerbuds are just appearing at the growing tip of the stem.

Left: A plant that is just starting to flower is not an ideal candidate for cutting material. The hormones in the growing tip will be directed towards flowering, rather than towards rooting.

1 *The ideal fuchsia cutting has a growing tip, one half-opened pair of leaves, one fully opened pair and about 0.5in (1.25cm) of clear stem.*

Always select cuttings from a good-quality, healthy plant.

The key to taking successful cuttings

Always cut the stem cleanly, using a sharp knife or similar tool. A blunt cut will damage the tender tissues. Never allow a cutting to dry out, as this will greatly reduce the chances of successful rooting. If necessary, place fresh cuttings in a small container of water. If you wish, you could add an extremely dilute solution of fungicide to the water to improve the plants' chances of survival. Use only the best plant material for cuttings. Remember, the better the cutting, the better the final plant.

The growing tip need not be present for the cutting to root. This internodal cutting will root and bush out without needing a first stop.

If cutting material is in very short supply, slice the cutting down the center. Use a sharp knife to minimize the damage; both parts should root well.

2 *Use a clean sharp knife to cut the stem. Make a straight cut that does not unduly damage the stem. A cutting with a damaged stem is more likely to damp-off than root.*

Cut just above a leaf joint. This will minimize the area of damage and effectively stop the plant.

3 *Hold the cutting by the leaf, rather than by the stem. Touching the stem can damage the fragile tissues of the young plant and cause problems with rooting later on.*

Ideally, avoid an uneven cutting, as the developing plant will also be uneven.

A three-leafed cutting makes an ideal plant, with three side shoots, rather than two.

Small cuttings root faster, as the growth hormone at the growing tip has less far to travel.

Raising the cuttings

It could be said that the most important part of growing fuchsias is taking a cutting, because from that initial piece of life, the most amazing plants can grow. Therefore, it is vital to produce the ideal environment for them to start. Fuchsias can be rooted in almost any medium, as long as it contains nothing harmful, but you must help them along by providing an atmosphere that will encourage rooting and growth. Propagators are available in many shapes and forms, from simple, homemade types to expensive, finely tuned commercial models. Whatever type you decide to use, they must provide three definite conditions: a humid atmosphere, the correct temperature, and an even level of light. These three factors are vital in most stages of a plant's growth, and a propagator should provide the right balance. Humidity is the most important: too dry an atmosphere will put the cutting under too much stress and it will not root; too wet, and the cuttings could damp-off, so you must aim for the right balance. Next in importance is the temperature. The majority of failures are probably caused by too much warmth; the optimum temperature is about 60°F (15°C). Finally light: the cuttings should root in an environment that is in light but not in direct sunlight; too bright, and the young cuttings will be under too much stress and fail. Once you have found the ideal growing environment for your cuttings, you should achieve success every time.

Homemade propagators

Homemade propagators can be made from a range of household objects and are cheap to produce. They must provide a humid microclimate for the young cuttings to root and grow. Each one performs like a mini-greenhouse that you can control, enabling you to water the cuttings and study them at will. Never let cuttings dry out - keep them cool and shady. Once they have rooted, open the container for a few hours each day so that they can become acclimatized to conditions in the outside world.

3 Use a fine stick or the end of a knife to make a small, shallow hole in the foam blocks to hold each cutting gently in position. Hold the cutting gently by the leaves, taking care not to damage the tender stem.

Use the cuttings as soon as possible after you take them.

1 Cut a block of flower foam into 1in(2.5cm) square blocks. This rooting medium retains water and creates a humid atmosphere.

2 Wet the blocks of flower foam thoroughly, not only watering them from above, but also letting them soak up water from the saucer. Keep them wet at all times.

Left: *A plastic bag holds a 5in(13cm) pot. Place the cuttings in previously wetted potting mixture and mist them well. Place a cane in the pot and secure the top of the bag to the stick.*

Below: *An old coffee jar over the potting mix will also create a greenhouse-type of environment. Be careful not to damage the delicate cuttings as you cover them.*

Below: *An old glass jar will take a tray with 8-10 cuttings. Its clear sides and screw top make it an ideal propagator, but heat can build up inside it, so keep it out of direct sun.*

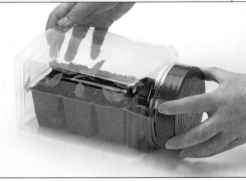

Above: *A seed tray filled with small pots or trays allows you to keep a variety of cuttings in one container. Be sure to label each batch clearly.*

Right: *You can monitor the progress of the cuttings through the clear plastic lid. The vents in the lid allow you to increase ventilation as required.*

4 *Gently insert the cuttings into the small holes. Do not force them in, as this could damage the stem, which is where the new roots will develop over the next 10-14 days. The roots will take up as much water as they need.*

The leaves should just clear the foam block.

5 *Cuttings rooted in this way can be placed directly into a suitable potting mix.*

Flower foam is an ideal rooting medium that contains nothing harmful.

Potting on and pinching out

When the young cuttings have rooted, the leaf tips become a much brighter shade of green and, without actually growing, the whole plant seems to elongate. However, this would be far too soon to move the cutting to its first pot; the average interval between taking the cutting and potting it on is one month. Weather, light and the time of year can all have an effect: it will be shorter in the spring when the sap is rising, but longer in the fall. Typically, a young plant should be about 2in(5cm) long before you pot it up. It should also have at least two fully grown pairs of leaves, plus the growing tip, and the roots should be well formed. The best time to carry out the first pinch out, or stop, is much the same as for the first pot. Do not leave it until the plant is too tall, unless you are thinking of growing it as a standard or another shape. Stopping performs two vital functions: firstly, it encourages the plant to become bushy and thus controls the shape. Secondly, at later stages of the plant's development, stopping gives you some control over the flowering time, so it is well worth mastering the technique early on.

Always plant into a small pot, as the shock of a move into too large a pot can slow down the plant's growth until it becomes established.

Left: When the cutting has grown to 1-2in(2.5-5cm), its roots should be well developed and it is ready to pot on. Remove the cutting from the propagator, holding it carefully so that you do not damage the tender young stem. Gently insert the cutting into the potting mixture.

The young plant is ready for its first pinch out when it has grown at least three pairs of leaves. Use fine scissors or tweezers to remove the smallest portion of growing tip. This allows side shoots to grow, forming the beginnings of a bushy plant.

How to produce a bushy plant

One stem stopped once will produce at least two stems, stopped again, at least four, stopped a third time, at least eight. In a cutting with three leaves at a node, the progression speeds up. A bushy plant is easy to achieve by pinching out after every two pairs of leaves.

This is a young plant before its first stop. At this stage, it has only one stem.

After one stop, the plant has produced a number of side shoots. They have developed from above the leaves on the main stem.

Each time you remove the growing tip on a stem, at least two branches will grow, and frequently more.

Getting your plants into shape

Stopping, shaping and turning are all aspects of growing fuchsias that tend to be overlooked in the desire to see them flower, but it is worth spending time on these tasks as the result will be manageable and pleasing plants. The importance of stopping a plant is explained on page 60-61. The number of pairs of leaves between stops is a matter of personal preference - a rampant grower will need stopping after every pair of leaves, while a slower, short-jointed variety would benefit from a looser style, so stopping after every two or three pairs of leaves would be better. The decision you make will control the final shape and style of the plant. The more you stop a plant, the bushier it becomes. It also lets you control the shape of the plant, which is essential if you are growing a standard or want to control a plant with wayward growth. Turning is critical to create a fuchsia with all-round growth and flowers. Ideally, the more you turn the plants, the rounder their shape will become. Aim to turn a plant 90° every three days, or more frequently if you have the time. Turning reduces the chance of a lop-sided, flat or triangular plant - a fuchsia that is not turned will soon grow towards the light. Turning your plants regularly helps you to get to know them as individuals, so that you notice when they need a little extra attention, perhaps a little more water.

This plant has not been well-stopped and shaped during its early stages. It is leaning over and lacks the ideal bushy shape.

Left: *At least twice a week, turn fuchsias 90° to ensure that they develop a well-balanced shape. The more you turn them, the more symmetrical their shape will be.*

This specimen of Ridestar has not been turned equally and has become triangular rather than round in shape.

This example of Waveney Gem has been turned equally and regularly and has a much more symmetrical growth pattern.

This plant has been stopped on a regular basis and is compact and bushy and also shows a nice symmetrical outline.

The right timing

To get your fuchsias in full flower when you wish, you need to calculate the interval between the final stop and when they are to flower. Singles need between eight and ten weeks, doubles between ten and twelve weeks. These are only guidelines, as the weather, light, temperature and feeding may all influence the timing. Each cultivar will vary slightly, but these guidelines will apply to most of them. If in doubt, always reckon on the longer time; it is far easier to remove flowers than to try and force a plant into flower.

Below: Waveney Gem in full flower. It will continue to bloom for many months. This delightful single-flowered fuchsia was photographed nine weeks after the final stop.

As a double, Ridestar is lagging a little behind; it will take somewhat longer to flower, as the blooms are that much larger. Most doubles take between ten and twelve weeks to give their best show.

Waveney Gem is beginning to show some large buds, five to six weeks after the final stop. Within two or three weeks, given good light, feeding and watering, it will be in full flower.

Below: Ridestar in full flower. It is a most delightful double with large blooms. Unlike many fuchsias with this coloring, Ridestar flowers fade very little once they have opened.

Potting up your fuchsias

Fuchsias benefit from being repotted. It is an opportunity to refresh the plant and provide it with a new growing medium. The plant will soon be encouraged to put on an extra spurt of growth. The 'pot-in-pot' method shown here is an ideal way of transferring a plant from one size pot to the next with the minimum of disturbance to the rootball, thus ensuring that any effect on growth is minimal. It also reduces any damage to the top growth, as you do not have to press the potting mixture down. There will be times when a move to a larger container is called for, such as when transferring a plant to a tub. In this case, it is probably easier to revert to the more normal type of potting technique. As usual, take care not to press down too hard on the potting mixture. Moving standards from pot to pot is another occasion to use the traditional method. The amount of potting mixture that you put at the bottom of the pot will determine the height that your plant goes into the new pot. This can be an ideal opportunity to lower a plant that is showing bare stems. You can put fresh soil on top of the existing rootball; this topdressing can be beneficial and encourages new growth.

3 Slowly lift the small pot out of the larger one. The soil should have formed a mold around the edges. If there is a gap or if it collapses, it is easy enough to start again.

1 Keep the difference in pot sizes to a minimum. If a plant is potted on into a pot that is too large, it will start producing roots to fill the new space before producing top growth.

2 Put a little potting mix in the base of the larger pot and place the smaller pot on top. Trickle soil gently between the two pots. Gently push the smaller pot outwards to bind the soil.

Wait until the rootball is well established and the plant is under minimal stress before potting on. For example, never pot on when a plant is over dry.

How to pot up your plants successfully

Do not pot up a plant until the roots are well established in the existing potting mixture. Never pot up a plant in full flower, as the shock could cause the plant to lose both flowers and buds. Never pot up a plant that is under stress, for example in full sun, too wet or too dry. Pot up and then place the plant in a cool, shady spot to recover. Never push the soil hard down into the pot. Modern, fibrous potting mixtures will find their own level. If they sink, just add more potting mixture to the surface of the pot. If the potting mixture is damp rather than wet it will fall into the gaps more easily. Varying the amount of mixture in the bottom of the pot enables you to lower or raise the plant in its new container. Dropping the plant in the pot allows you to hide any bare lower stems. To achieve the effect of an extra large plant, try potting three similarly sized plants into one large pot. This is an ideal way of filling large tubs and containers.

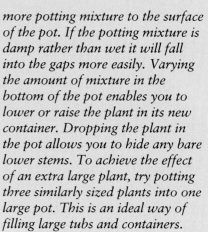

5 A plant that was ready for potting on always looks at home in the larger pot. Potting up by this method ensures that the plant continues growing with the minimum delay.

4 Carefully lower the plant into the mold that you have made - a steady hand is useful! Once it is fully down, add a little fresh potting mix to the surface of the rootball. This will encourage the plant into even more new growth.

After potting, water the plant gently and keep it in a cool, shady position while it recovers from the upheaval. If the level of the soil sinks, add a little more.

Always use clean pots. If necessary, wash them thoroughly before use. Failure to clean the pot can lead to the spread of soil-borne pests and diseases.

Root trimming a fuchsia

There will be times during the growing season when you would like to keep a plant in the same size pot, but the plant has become potbound and the foliage is beginning to look stale. This is a time for drastic action! If you remove about 1in(2.5cm) from the base of the rootball and put the plant back into the same pot, you will achieve two things. Firstly, you will refresh the plant by giving it a new root run; after a small check, the plant will again grow briskly with fresh and healthy growth. Secondly, you will be able to drop the plant slightly lower in the pot, which will allow you to put some fresh potting mix on top and cover up areas of bare stems where the lower leaves have dropped off. If it will not drop lower in the pot, shave a little off the sides of the rootball so that it will fit snugly. The fresh potting mix on the top will also promote healthy growth, as the feeding roots are found there. This fresh layer of soil will soon be full of roots and the plant will race ahead of those that have not had this treatment.

2 *Begin to cut the base of the rootball. Never carry out this action when the plant is in full flower, as the shock could cause it to lose its blooms. A few lost buds will not delay flowering by much.*

Below: To keep a plant such as this in full flower for the whole summer, you may need to take decisive action! Lucinda, a fine semi-double, will flower profusely in good conditions.

1 *This Waveney Gem fuchsia has been in the same size pot for many months. Its roots are beginning to circle the pot - a sign that it is becoming potbound.*

Use a sharp knife or a saw if the plant has large tough roots.

Do not be afraid to tackle this procedure, as fuchsias replace their roots very quickly. With a large, old plant it may need doing every year.

3 Cut a clean slice about 1in (2.5cm) across the base of the rootball and gently remove it. If necessary, also slice a little soil off the sides of the rootball.

4 Add an equivalent amount of fresh soil to the base of the pot. It is better to use too much than to leave an air gap that will not help the plant.

5 Lower the plant into the pot. It should rest gently on the fresh soil. If it feels very tight against the sides of the pot, remove it again and slice a little off the sides of the rootball.

6 Press down lightly on the rootball to remove any air spaces and add a little more fresh soil over the surface of the existing rootball. Water the plant gently to settle the soil down.

7 Four weeks after this apparently drastic treatment, Waveney Gem is now full of flower. Following the addition of the fresh potting mixture, it is looking healthy and refreshed.

Hybridizing fuchsias

There can be nothing more fascinating than the prospect of producing a brand new cultivar by crossing different fuchsias. However, be warned; there are approximately 10,000 fuchsia varieties around the world, so the chances of producing something entirely different are fairly small! The photographs on these pages clearly show how to carry out the fertilization process. Once the seedpods have matured and are looking like little grapes, remove them carefully from the plant. Cut each fruit open and you will see a central area surrounded by a mass of small seeds. Detach these using a cocktail stick or similar implement and tip them onto an absorbent surface. Viable seeds are larger than the rest and often darker - they will also sink when placed in water. Dry the seeds and keep them safe and carefully labeled. Plant the seeds at the normal time and in the usual way. Be patient, as some may take quite a while to germinate. Do not throw away the potting mixture for several months as something special may still appear! Grow the plants as you would any other fuchsia. Waiting for those first flowers to open can be quite an experience! Take cuttings of any plant you consider good enough to keep and nurture them for another three years to ensure that the cultivar holds all its characteristics. Your local specialist nursery or fuchsia society will advise you if it is worth continuing with your new cultivar.

Pop open an almost mature bud by gently squeezing it so that nothing is damaged inside.

1 *Choose a cultivar with characteristics that you like and prepare it to become the female partner in the pollination process. Selecting a young bud is vital so that you can isolate the female parts before they become mature and open to pollination.*

Fuchsias have eight stamens and they must all be removed. At this early stage they can be well hidden in the petals, so hunt around if they are not immediately apparent.

2 *Use sharp scissors to cut away all the anthers and filaments (male parts of the flower). This will prevent the flower pollinating itself. Be careful not to damage the flower.*

Twist ties are fine for sealing the bag, provided you use them carefully.

3 *Gently enclose the emasculated flower in a plastic bag. This will allow you to see the flower as it matures, but also to give the flower protection from being pollinated by another fuchsia.*

4 Test the maturity of the flower by placing your finger gently on the tip of the stigma - if slightly sticky then it is receptive. To make the rest of the procedure easier you can remove the petals if you wish.

5 Carefully remove the pistil (female parts) from the flower chosen to be the male one. Test for ripe pollen by brushing your finger over it - ripe pollen grains will stick to your finger.

6 With both flowers ready, remove the plastic bag from around the female flower and bring the male flower to it so that the pollen grains can be accepted by the stigma.

Gently brush the two flowers together. Keep a note of the cultivars used and the date, particularly if you do a number of crosses between the same or different cultivars.

7 After the pollination process, isolate the flower again to avert random pollination if your attempt has failed. Use a plastic or muslin bag to protect the flower while the seedpods develop at the base.

You can remove the bag once the flower has fallen off, leaving the seedpod.

A young pair of seedpods. Fuchsias have either green or reddish purple pods at maturity - which can take several weeks.

8 It is possible that the seedpods (berries) may fall off at an early stage, indicating that fertilization was not successful. Therefore, it is worth doing several flowers at the same time to increase your chances of success.

New styles in fuchsia flowers

One of the first great color breakthroughs in the world of fuchsias, and quite a dramatic one at the time, was the production of the first white tube and sepals found in Venus Vitrix, which was introduced in 1840. It is still a most unusual flower. Following on from this came the introduction of orange flowers and many variations on existing colors, which have resulted in some incredible combinations of color, size and shape. It was not until the 1980s that hybridizers in Holland tried all kinds of crosses between species and triphyllas, and species and cultivars. Amazing things started to happen and the claret and burgundy-colored fuchsias appeared. These have opened a whole new world of color that we are only just starting to explore. Hybridizers are still searching for the truly yellow flower. *F. procumbens* from New Zealand has yellow within its flower, but it is genetically so far removed from the rest of the fuchsias that it is virtually impossible to cross it with any other flower. Over the years, several cultivars have appeared with yellow in them, but they have never been released as they have always been weak growers or had other problems. Time will tell whether the dream of a yellow flower is ever realized and whether we will like it when we find it. Some hybridists are on the trail of other color combinations - the pure orange double, for example. All the current ones have a fair degree of pink in them, as well as orange. Others are hunting for a flower with blue petals that do not fade to a washed-out lilac color. There is certainly plenty to keep the hybridists occupied in the years to come.

Above: Haute Cuisine is a good-sized burgundy double, and ideal for a hanging basket. It opens as an incredibly dark flower and fades to an attractive lighter shade.

Right: Rina Felix is one of the more unusual crosses. The shape of the flowers and the growth habit have many triphylla characteristics. However, the attractive plum flowers are a new color to the triphyllas.

If you examine the petals closely, you will see that they are almost diamond-shaped.

Mood Indigo has an extraordinary growth habit. It is one of the most rampant growers and will fill a hanging basket with great ease.

Below: *Fuchsiade 88 was one of the first burgundy singles to be released. It has an unusually stiff, upright growth and an abundance of flowers. It also makes an excellent hardy planted in the garden, as it is remarkably tough.*

The angle at which the sepals are held varies according to the individual cultivar.

Left: Mood Indigo is perhaps the most attractive of the burgundy fuchsias, as the white makes a delightful contrast with the unusual color of the petals.

1 Remove any damaged or marked blooms from the plant. These show up clearly on a light-colored plant such as this one, but on a dark-colored fuchsia look for flowers with brown or discolored areas.

Showing fuchsias

Once you have grown a plant of which you are justifiably proud, you may well feel that the time has come to show it to everyone else! Most specialist societies have an annual show or display that could provide you with just the opportunity. Taking part in such an event, whether it is competitive or not, is great fun and allows you to meet other fuchsia enthusiasts in a friendly atmosphere, and the organizers are always keen to have as many plants as possible to make the occasion a spectacle. Any plant that you consider taking along must be pest- and disease-free and have a good, healthy appearance. Ideally, the plants should be symmetrical and full of flower, the flowers being typical for the cultivar. There will be opportunities to include most types of fuchsias, and often there is an area where individual blooms are exhibited, either for artistic effect in a shallow dish, in sand or in small containers for six or twelve blooms. Each bloom should be as perfect as you can find, typical in shape, size and color and with no markings on it. The anther and filaments and the stigma and style should all be fully extended. Pollen should just about be showing on the anthers. Get all these factors together and you will have the perfect flower!

2 Remove the green seedpods right back to the main stem, as any small pieces of stalk will detract from the final effect.

3 Search carefully for any yellow leaves and remove them as well. Remove any debris, such as leaves and dead flowers, from the top of the potting mixture.

4 Show plants need to be as symmetrical as possible and often need a final tie-in just before the show. If the plant has a lax growth, consider tying it in at the time of the last stop.

5 Insert three canes into the potting mixture in a triangular pattern around the edge of the pot. With a particularly large plant, you may need to use four canes.

6 Draw a string between the canes, bringing the plant in to create a good shape. Tie any loose branches separately to a cane or to another branch. The string should not be seen.

Olive Moon ready for the show. This type of clearing up will help the plant to continue in flower for many months.

7 Using sharp secateurs, cut off any protruding pieces of cane, so that they are below the level of the majority of the growth. Do not cut the plant!

8 Once you are happy that the plant looks its best, wipe the pot with a damp cloth. Position the plant with its best side facing forward and leave the rest to the judges!

Part Two

DISPLAYING FUCHSIAS

Fuchsias in full bloom put on a superb display. It is not surprising, therefore, that these graceful flowers are extremely popular around the world for the color and elegance they bring to containers and garden borders alike. Fuchsias have so much to offer. There are pastel shades that blend harmoniously or vivid oranges and reds that will brighten the darkest corner. There are single-flowered plants that are almost constantly in bloom and exotic cultivars laden with fewer but larger double flowers in truly wonderful color combinations. And, of course, many fuchsias have bright or variegated foliage to take the eye.

This section of the book presents a wealth of display ideas that highlight the beauty and versatility of fuchsias. After a brief reflection on the charms of old varieties, attention turns to fuchsias in summer bedding schemes, containers and mixed hanging baskets. Growing fuchsias into fans, pillars, hoops, spirals and other shapes has been popular since Victorian times and the necessary growing and shaping techniques are fully covered here. This is followed by a fascinating look at growing fuchsias in unusual containers, how to raise them as bonsai specimens and simple ways to maintain them successfully indoors. The section closes with several flower arrangement ideas using cut fuchsia blooms. Although this is relatively uncommon, fuchsias can make stunning indoor displays, whether as individual blooms floating in water or as cut stems mixed with other flowers to make a traditional table decoration.

Left: *Fuchsias grace a mixed flower display.* **Right:** *The exquisitely tinted blooms of Vanessa Jackson.*

Discovering old varieties

Left: Lye's Unique was introduced in 1886 by James Lye, one of the great hybridists. The waxy white of the tube and sepals was very much his trademark. It is still popular today.

Fuchsias were first introduced into Europe in the 1780s. Legend has it that a nurseryman, Mr Lee, spotted a fuchsia growing in the window of a sailor's house in Wapping, London. He recognized it as a plant with sales appeal and, having built up a stock of some 300 plants, was soon doing a brisk trade. Since then, the fuchsia has rarely looked back and there cannot be many places where it is not found today. Many cultivars from the 1800s have tremendous strength and vigor and have survived in cultivation to this day. There is every reason to believe that people will still be growing them in 50 or 100 years time, when many of the more modern cultivars may have faded away. They must be something quite special to have survived this long and are well worth growing.

There is a tendency to believe that color breakthroughs, such as orange, are new, but they first appeared in the 1880s and are still with us today. The early hybridizers experimented more and more as the new species were discovered in South America.

Left: Bland's New Striped was ahead of its time and looks very much like a modern introduction. In fact, it was first seen in 1872 and is still flourishing and in demand today.

The foliage of Tom West will become more pink in the sun. A change of feed also often alters the shade of the foliage.

Above: *Venus Vitrix was the first fuchsia to be found with a white tube and sepals. It is worth persevering with this small, but sometimes temperamental, plant because of the striking combination of white and almost blue flowers.*

Left: *Tom West, an excellent hardy, is still a very popular foliage fuchsia, but was first found as a sport, possibly from Fuchsia Corallina, in 1853. It will enhance any modern collection.*

Old varieties to look out for

Bush: *Bon Accorde (1861)*
Clair de Lune (1880)
Grus aus den Bodenthal (1838)
Lye's Own (1871)
Lye's Unique (1886)

Hardies: *Avalanche (1869)*
Chillerton Beauty (1847)
Herald (1847)
Madame Cornelissen (1860)
Snowcap (1888)

Summer bedding

Using fuchsias as a form of summer bedding, rather than as permanent planting, can add a whole new dimension to your garden and gives you more scope to experiment. Use standards as part of your planting scheme to increase the range of possible heights. Add further contrasts by using the larger doubles and triphyllas, but bear in mind that neither of these should be left outside for the winter months, although they will do very well during the summer. Look for good color schemes to go with your other bedding plants, either to blend in with colors you already use or to add a contrast. For example, fuchsias can add a great deal to an orange and white area or to a pale pink one. When planting areas of summer bedding, why not try groups of three or five plants of a cultivar in one area - odd numbers always look best. Group plantings will maximize any effect that you are trying to achieve. Be even braver and try planting a whole bed with just one cultivar. There is nothing more eye-catching than fuchsia flowers en masse. Fuchsias planted out for the summer will still need a shady spot and regular watering, as their roots will not have had time to establish a good root run. Do not be tempted to leave them in the garden in their pots. If it is very hot, the roots will burn and their ability to take up any available water will be limited by the drainage holes in the pot. Regular feeding is vital to maintain continual flowering. Once a week, apply a balanced feed with a ratio of 20:20:20 (see page 40). This will increase the fuchsias' natural desire to flower even more. Be adventurous and beautify your garden with fuchsias.

Left: For a really eye-catching splash of color, combine bright fuchsias with a selection of summer bedding. Here, pelargoniums, ageratums and begonias are set off by the silver foliage of Cineraria maritima. *Standards add height and depth to the display.*

Below: *Although it is a double, Paula Jane has a rigid growth habit and holds its vivid ruby colored flowers aloft. It will brighten up the garden, particularly when it is grown alongside impatiens and lobelia.*

Right: *Here, a traditional red-and-purple fuchsia is teamed with a dark heliotrope to provide a reasonably restrained color scheme. Use colors to either brighten or calm an area of the garden.*

Below: *Fuchsia paniculata is a tender species and an unusual, but effective, choice for summer bedding. The large grapelike fruits that appear after flowering are very popular with birds.*

Tubs, troughs and windowboxes

Fuchsias are ideal plants for any kind of container, be it a tub, a trough or a windowbox. Whether you plant them on their own or with other plants, such as ivy-leaved pelargoniums and brachycome,

Right: The plant and container should be in proportion to one another. This specimen of Thalia suits the fairly tall half-barrel and does not mind a warm, sunny spot.

they will flower continuously for many months. To keep them looking their best, carry out regular general summer maintenance, which means removing dead flowers, seedpods, etc. Use their delightful range of flower colors to their best advantage. Try foliage fuchsias to increase your range of possibilities even further, as they will provide a splash of color long before they are in flower. When planting up the container, check the drainage, as a waterlogged fuchsia will soon suffer and possibly die. If drainage is poor, improve it by adding grit, pieces of broken flowerpot or small pebbles to the bottom of the container. As with baskets, add a slow-release fertilizer at planting time. This can save you having to remember to feed your fuchsias regularly, unless you feel that the plants need a boost. Choose plants and colors that will suit you, mixing uprights and trailing varieties, if the situation will allow. If not, keep to one type only. The colors of plants never seem to clash, so let your imagination run riot!

Right: A nicely balanced windowbox, featuring a strong-growing, upright double fuchsia with a selection of trailing plants - Brachycome and Nepeta - to provide a contrast of shapes and textures.

Left: *This delightful feature at the corner of a wall makes a focal point for the garden. To succeed, the color of the fuchsia must complement that of the other plants and the container.*

Make certain that the container provides ample drainage for the plants.

Left: *For something really exotic with a tropical look, try planting up a large container with a good-sized specimen of Thalia, plus Cordyline and some pelargoniums for added color.*

Above: *Any kind of container can look good in the garden, as long as it has ample drainage holes. Here, a fuchsia is clearly thriving in a classic chimney pot with an open base.*

Using fuchsias in mixed hanging baskets

When you look up at fuchsia flowers, you are seeing them as they should be viewed, so when they are planted in hanging baskets you can really appreciate the blooms. The large doubles do superbly well in baskets, as the weight of their flowers pulls the stems downwards. Mixed baskets open up a range of possibilities, because incorporating other plants ensures a longer flowering period than a basket of fuchsias alone can offer. The mixture of plants can be very individual, allowing you to choose between, say, bright colors or a more subtle color scheme. However, it is important to achieve the right balance of plants. For example, do not combine sun- and shade-loving plants, as one group is bound to suffer. In a mixed basket containing a number of fuchsias, select plants that enjoy a degree of shade. Modern baskets often have holes in the sides for extra plants, but do not be tempted to pack the basket too tightly, as all the plants will need space to grow. Remember that a mixed basket needs plenty of regular feeding, as the plants will all be competing for nutrients.

Use foliage plants to enhance the effect of a mixed basket. They will give color to the display before the basket is in flower.

Left: Mixed baskets provide a wide variety of colors and textures. This one contains eight different types of plants, which means that some of them will always be in flower.

Right: This basket catches the glow of the sandstone wall. It contains a variety of foliage plants and Marinka, a rich red basket cultivar that flowers continuously for many months.

Below: *Generally speaking, it is a good idea to plant just one cultivar in each basket. However, a mixture of fuchsias, combined with other plants, works successfully in this display.*

Above: *Try color contrasts, as well as harmonizing color schemes. Choose a scheme that suits you and the planting situation.*

Above: *The depth of color and the size of its flowers make Swingtime, in the foreground, one of the most popular basket cultivars. Single cultivar baskets hanging together can make a striking display.*

Creating a fan of fuchsias

The Victorians were perhaps the finest exponents of this form of growing fuchsias. The oldest fuchsia books feature photographs of incredibly tall plants and show how they were carried around on carts from place to place. However, do not let this deter you. Fans and pillars need not be tall; smaller ones are much more manageable and highly acceptable. Coincidentally, many of the older fuchsia varieties, particularly those considered as hardies, are some of the best for growing as pillars or fans. Once the structure is established, it can be maintained for many years. In fact, the plants really mature and are often much better in their second or third season. When growing a fan, it is very much a case of looking for the right cutting, namely one that has a definite spread to it and looks as if it can be trained to the appropriate style. Stopping is again particularly important to determine the plant's shape. Pinch out the side shoots to add depth and solidity to the framework of branches. Always aim for a symmetrical, well-balanced shape. Espaliers are a similar form of growth, with one central stem and the laterals trained horizontally. You can have fun experimenting with different shapes if you find the right cutting to start with.

5 Use a plant tie or a soft piece of string to secure the plant to the cane. Do not tie it too tightly. Keep the growth as straight as possible.

4 Gently direct the growth towards the cane, without damaging the stem. Carry out all positioning before the growth becomes hard and woody.

1 Choose a plant of a cultivar that you know to be a strong grower. Ideally, it will have had one pinch out and have produced at least three strong-growing branches.

2 Insert small canes carefully into the rootball. Do not replace them until a later stage when the fan is considerably larger. Space the canes symmetrical within the pot.

3 Fasten another cane across the top to give the structure greater stability. If necessary, add a second cane lower down, particularly as the structure gets larger.

6 *You can produce fans in all kinds of shapes and sizes. Ideally, an odd number of branches will give a more pleasing shape once they have been trained.*

7 *To create a good structure, begin stopping as the plant reaches the top of the first set of canes.*

8 *Stopping the top growing tips will cause the plant to bush out. Doing the same on the developing side shoots will result in a plant with a more solid structure.*

The foliage on a fan should fall right down to the pot, so that the branches are well hidden.

With its striking flowers and strong growth, Banstead Bell is an ideal cultivar for a fan.

9 *A small fan during its first season of growth. In the course of the following season the growth will come from the old wood and the plant will be much larger than before.*

1 *To start training a pillar, choose a plant that has one strong leader, preferably with a number of side shoots. Insert a cane through the center, being careful not to damage the main stem. You can always use a larger cane later on if this one seems too small or insubstantial.*

2 *Tie the main stem gently to the central cane. It is important not to damage the tender young growth at this early stage of the process.*

Creating a pillar

Pillars are perhaps easier to grow than fans, as they are very much like a standard with side shoots left on all the way down the stem. The stopping program is again important, as is the use of canes - both will help to produce a sound structure. When it is finished, a pillar should be symmetrical and the same width across the top as it is at the base.

Another shape you might like to try is a pyramid, i.e. a plant that is broader across the base and coming to a point at the top. Once again, the basis is much as a standard and careful, regular stopping will produce a lovely and unusual triangular plant. This is not quite as easy to achieve and needs some practice, so try growing pillars and fans first and then attempt some of the other shapes later on. It is a challenge to tackle these old ways of growing fuchsias, but it pays dividends!

Try to keep the growth as straight as possible.

3 *Start to improve the shape by pulling the side stems into the center. Use a soft tie to fasten them in. Do this while the stems are young and still malleable.*

Do not pot up the fuchsia until the roots are starting to fill the container.

4 *Once you are happy with the form of the plant, pinch out the side shoots to make the shape more bushy. Leave the main central stem unstopped, so that it continues to grow taller to produce the pillar.*

Pinch out the side shoots, but do not remove too much growth. The aim of the exercise is to improve the shape of the plant.

Right: *A large pillar of* F. magellanica aurea *just coming into flower. Try putting more than one plant in a pot like this one to achieve a large pillar more quickly.*

5 *Keep on tying in side shoots until you are happy with the height and shape of the plant. Then discontinue stopping and allow the plant to flower.*

Frames and hoops

Fuchsia growing should be fun and it is not always important that the plants are grown in conventional shapes. There are times when it is worthwhile experimenting. Many cultivars have a very supple growth that you can bring under control with a little training. The advantage of growing fuchsias around wire or plastic shapes is that you do not have to worry about stopping times - as the plant grows, it flowers! There are several points to bear in mind. Firstly, if you are using wire, try to use a plastic-coated kind, otherwise the heat from the wire can burn the plants during the summer months. Secondly, think about the cultivar that you are choosing. If you want a small shape, use a small-flowered type, as a large flower would look totally out of proportion. It is not always necessary to use the most floriferous cultivars, as continually tying or bending in the young growth will enhance any floral display simply because the flowers are so much closer together. This applies particularly to the encliandras; with their extremely small flowers they can look sparse in other circumstances. Using more than one plant in the pot can make growing a three-dimensional structure much easier.

Finally, do not be discouraged if you do not achieve the expected result first time; a little perseverance will certainly pay off!

Decorative shapes

Making shapes is as simple as bending the wire. As long as the angles are not too extreme, fuchsias will accommodate to a wide range of forms. Why not try your house number or your initials? All it needs are nimble fingers and a degree of patience. Be imaginative and adventurous!

Right: *F. hidalgensis, with its attractive, small white flowers, has been used to form this globe.*

Far right: *A spiral is easy to make using the same species, one of the most supple for this purpose.*

Cultivars for growing into shapes

F. hidalgensis, F. hemsleyana, *La Campanella, Lottie Hobbie,* Mrs. Lovell Swisher, Mrs. Popple, Neapolitan, Northway, Pink Rain, String of Pearls, Topper, Whiteknights Pearl

Use a plastic frame or plastic-covered wire to avoid the risk of scorching growth in hot weather.

1 *Fuchsias can easily be trained into a shape while they are young and supple, so start at this stage. Add the wire before the plant has grown too large, shaping it to the required size and shape.*

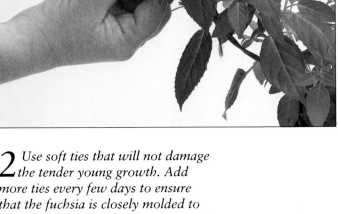

2 *Use soft ties that will not damage the tender young growth. Add more ties every few days to ensure that the fuchsia is closely molded to the wire shape while the growth is still supple. Fuchsias can grow quickly!*

Coping with large shaped fuchsias

Take care when potting on
fuchsias that have been
grown in elaborate shapes.
You may need a helper to
balance and support the frame.
Large hoops are a little flimsy until they
have become established, so do not leave
them in a windy position. Be sure to give
these pots a full half turn to ensure that
they grow evenly; a quarter turn is not
enough. For the best results, tie in
the growth every two or three
days. This ensures that the
shape remains neat.

3 The plants need
no stopping,
because they flower as they
grow. It is important to
maintain a regular feeding
program to ensure fine
healthy growth at all times.

4 Eventually the
circle is complete
and Whiteknights
Pearl is beginning to
come into flower.
Choose a cultivar with
flowers in proportion
to the size of the
shape. A strong
grower is also useful.

Fuchsias in novel containers

Let your imagination run riot when considering the containers in which your fuchsias will look good growing and flowering. They do not have to be on a grand scale, but make sure they are in proportion to the size of the plant and the flowers. Small plants in small containers can be great fun. Look for containers with some kind of drainage. If there is none, then it may be possible to make holes in the base, but only if you do not want to use it for anything else afterwards! If drainage remains poor, water very carefully, giving your plants just the merest drop as necessary. A terracotta container with poor drainage is not such a problem as it is porous and water can seep out of the sides, thus reducing the potential problems. Try matching fuchsia names with containers; for example, Trumpeter in a trumpet, Peppermint Stick in a candy jar. Or why not match the size and shape of the plant to the style of the container?

If possible, put the plant in your chosen container long before it is in flower so that the stress of the change does not cause it to drop flowers and buds. If it has to be a last-minute enterprise, keep the plant in a cool, shady spot for as long as possible to reduce the stress and allow it to recover from the change.

Points to watch

Take care when watering novelty containers, as many have poor drainage or none at all. Water them sparingly and do not forget to feed them. Try to find containers that are in proportion to the size of the flowers and growth habit of the plant; a small plant in a large pot (or vice-versa) can look very strange.

With its slightly trailing growth, Autumnale (Burning Bush), makes an ideal plant for this small display. As the plants are in ordinary pots on saucers it is easy to change the plants.

A watering can is an ideal container, provided you make plenty of drainage holes in the base. Do not put it in full sun for long, as the heat could cause the roots to scorch. Wassernymph (Waternymph) seems an ideal choice.

Dancing Flame makes a fine display trailing over a terracotta cauldron. The plant can lose moisture through the porous pot, as well as via the drainage holes.

A plant in a shoe is always a talking point. Push the potting mix very gently to the extremities, so that the roots can spread around. Olive Smith, a small, single-flowered cultivar, seems to be thriving.

A small kettle with a large plant; this one is Superstar. Take care when planting around the handle. Gently maneuver the branches around it to achieve the desired effect.

Drainage can be a major problem in a glazed pottery teapot with no holes. Water very carefully, giving only small quantities as and when the plant and potting mixture become dry.

1 *Remove as much soil and as many fibrous roots as possible from the plant you have chosen to train, so that it will fit into the shallow bonsai container.*

2 *If there are any larger roots that would be difficult to fit into a shallow pot, carefully trim these with secateurs. Do not worry - your plant will survive.*

Fuchsias as bonsai

Growing fuchsias as bonsai subjects is as much of a challenge as creating large structures. However, fuchsias are more than happy to be root-pruned and this makes them ideal candidates for bonsai. The most important thing is to choose the right varieties of fuchsia to grow, as the proportion of the plant to the container and the fuchsia's general growth pattern are clearly vital considerations. Large flowers on a small plant in a small pot would look quite wrong. You will have to rethink some basic gardening principles, too. For example, a plant with a balanced shape is no longer the aim. Have a look at any less than perfect plants and you may well find a potential bonsai. Plants with a woody stem in their second year are ideal candidates, and as with any other bonsai, you can use wire to help create a shape. Judicious pruning and positioning in the pot make the plant look more authentic.

Never give bonsai fuchsia a high-nitrogen feed; a weak solution of a balanced feed is best. Keep the plants in a shady position to prevent them drying out. Gentle watering with a fine spray will gradually bring the top roots to the surface and expose them, and you can further enhance the bonsai effect by adding a little moss. Bonsai fuchsias can remain in the same container for many years. Give them an annual root prune and add some fresh potting mixture at the same time. They will give you much pleasure in this unusual form.

3 *Choose the best position for planting your fuchsia in its pot. It need not be symmetrical, but should create a pleasing, bonsai-like appearance.*

4 *Gently trickle more soil into the tiny gaps around the roots. A small spoon is ideal for this. Do not press the soil down; it will find its own level.*

Put the moss gently in position. It will soon regrow.

5 For a final bonsai touch, add a small piece of moss around the main stem. Put it in place and shape it as necessary. This piece came from the surface of an old fuchsia.

6 This is a good opportunity to trim branches and carry out any stopping required to shape the plant. To achieve an authentic bonsai shape, wire soft branches into 'windswept' positions.

7 Once you are happy with the plant, give it a little water and place it in a cool, shady spot. Leave it in this stress-free environment for up to a week and water it as necessary.

Suitable fuchsias to grow as bonsai

Where possible, try to look for plants with small flowers and a small growth habit, as they will look ideal as a bonsai. With their small leaves and delicate flowers, encliandras are good subjects, and so are the dwarf-growing hardies, such as Tom Thumb and Lady Thumb. Try foliage fuchsias as well. Always be prepared to experiment.

A three-year-old Pumilla fuchsia.

David, an old, low-growing hardy, is an excellent choice.

To achieve this sort of shape, bend wire gently around a young and supple stem.

93

Growing fuchsias indoors

Fuchsias do not always make ideal houseplants, but given a little extra care and attention they will thrive. Fuchsias need a humid microclimate all around them, so spray them every day if possible, but be careful of the furniture! You should also take them once a week to an area where you can mist them thoroughly to clean them and remove dust and grime from the leaf surfaces. Never place fuchsias too close to direct sunlight or to a source of heat. Make sure that the plants are not stressed, particularly if they are in flower or bud. Stress at such a time will cause them to lose both buds and flowers. If possible, start growing the fuchsia indoors as a small cutting and keep it inside so that it becomes accustomed to the environment. This is safer than expecting an outdoor plant to adjust to a sudden change. Feeding and general care are just as important indoors, as is vigilance regarding pests and diseases. Turning is perhaps even more important indoors, as the light will be coming from one direction. Give the plant a quarter turn once a day to achieve the best all-round appearance. With care and attention, it is possible to grow any fuchsia indoors, but it is a good idea to concentrate on short-jointed and compact plants, as the others tend to become straggly and rather unwieldy indoors.

1 Choose a plant that is not in flower. Put some small pebbles into a large saucer. This will raise the plant above the eventual water level so that it does not become waterlogged.

2 Add some water to the saucer, taking care that the level does not rise above the pebbles. The plant should not sit directly in water for long periods of time.

3 Put the plant onto the pebbles. Water evaporating from the surface of the pebbles will provide a humid microclimate for the plant.

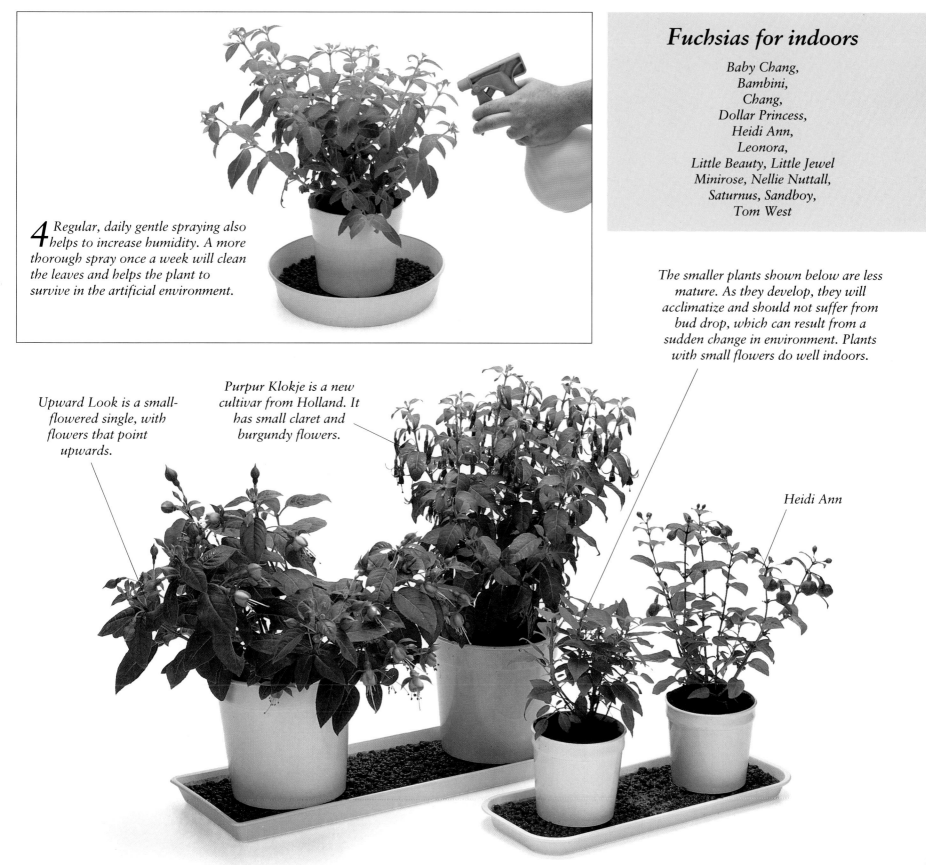

4 *Regular, daily gentle spraying also helps to increase humidity. A more thorough spray once a week will clean the leaves and helps the plant to survive in the artificial environment.*

Fuchsias for indoors

*Baby Chang,
Bambini,
Chang,
Dollar Princess,
Heidi Ann,
Leonora,
Little Beauty, Little Jewel
Minirose, Nellie Nuttall,
Saturnus, Sandboy,
Tom West*

The smaller plants shown below are less mature. As they develop, they will acclimatize and should not suffer from bud drop, which can result from a sudden change in environment. Plants with small flowers do well indoors.

Upward Look is a small-flowered single, with flowers that point upwards.

Purpur Klokje is a new cultivar from Holland. It has small claret and burgundy flowers.

Heidi Ann

1 *Before mixing the gel with water, read the manufacturer's instructions carefully and check the proportion of gel to water. Put a spoonful of gel into a large bowl.*

Fuchsia heads in a blue glass bowl with gel

Displaying fuchsias as individual blooms gives you the opportunity to discover the beauty and complexity of each flower. You can show off your favorite varieties and pick the most colorful combinations to make a spectacular table centerpiece. When a bloom is upturned, it is possible to see clearly how each flower is constructed and just how pretty the color combinations of petals are; it is as if you had a completely new species of flower to use. If you cut off the stem just below the lower petals, each bloom will last very well as a cut flower, even without any stalk. Arranging and displaying the fuchsias in a flower gel gives them stability and the best conditions for lasting. The gel is sold in the form of dry crystals and swells dramatically when water is added. It makes a superb medium for this kind of arrangement, holding the bloom in place and providing water to the cut stem. You can even add food coloring to it to create different and exotic effects without harming the flowers.

2 *Add the required quantity of clean, cold water and stir well for a few seconds. Now leave the gel to absorb the water to its maximum capacity.*

3 *The gel will be ready to use after about ten minutes. Tip most of it into a large shallow bowl and the rest into each small glass.*

4 *Prepare some single blooms from the stems of fuchsias. Cut each flower off just where the stem finishes under the head. Choose perfect specimens that are just fully opened.*

5 *Place the fuchsia flowers into the bowl, arranging them into the gel which will support them in an upright position. Leave space around each bloom. These are Reading Show.*

Marinka

Mixing several varieties together makes the most impact in such a simple arrangement.

Seeing the fuchsias in profile sheds a whole new light on them as flowers.

Rufus

Ben's Ruby

Billy Green

6 *Arrange the two small glasses in the same way. The finished bowl looks superb, especially as the strong color of the glass contrasts with the rich reds and purples of the fuchsias.*

Mixed fuchsias in a moss-lined basket

This is a lovely way to display different fuchsia varieties together in a fresh and informal arrangement. The green wire basket and moss look just right with the arching sprays of fuchsia. The fuchsias used here are all quite similar types, with medium to small flowers, both single and double. Basket types are suitable, as well as upright-growing kinds. Cut them from large plants and leave the foliage intact to provide color contrast and definition. It would be an ideal way to display branches of hardy types that you have picked from the garden and brought indoors. Treat the stems of fuchsias as you would any other cut flower material. Split woody stems a little to allow better absorption of water and cut green stems at an angle neatly and cleanly. You could adapt this idea and create it using a basket made from other material, although the moss is an important part of the design and should be clearly visible through the sides of the container.

5 Start to push the fuchsia stems into the foam. The aim is to create a pleasing all-round arrangement that mixes the varieties fairly randomly.

1 Start by lining the wire basket with fresh, damp moss. You will find it easier to work using several small pieces rather than struggling with one unwieldy mass.

2 Push the pieces of moss against the sides of the basket and as far up as possible while the basket remains empty.

3 Now line the basket with a piece of thin black plastic cut roughly to shape. This will hold in the moisture later on.

4 Cut a block of damp flower foam to fit inside the basket, stopping just below the top of the basket. Put it into position.

6 Continue building up the arrangement, adding more stems to make a fairly dense mixture. Let one or two stems drop down low at the front and sides of the basket.

Snowcap

Pacquesa

Dollar Princess

Margaret Brown

Rufus

Mary

7 Stand the finished basket either at table height or a little higher to appreciate the hanging fuchsia flowers better. Spray occasionally with a fine mist of water to keep the basket fresh.

99

Fuchsias, pinks and lilies in three-tiered glass

This is a classic and very glamorous arrangement for a special occasion. Although it looks grand and formal, it is exceptionally easy to put together. You will need two cake stands that balance on one another and both should be just deep enough to hold a little water for the flower stems. The lilies and pinks create a good foil for the more delicately shaped fuchsia flowers that tumble downwards from the display. The whole idea is reminiscent of designs used in Victorian times to decorate buffets and dinner tables, as fuchsias were very popular during that period. If possible, construct the arrangement in situ, as it is very difficult to move once it is made up. You could try a different version of this idea using a thin layer of damp moss on each stand and setting the flowers amongst it. This would work well if the stands were made of china rather than glass or where they are too shallow to hold enough water.

3 Take a stem of lilies with two or three blooms and cut the stem short. Lay it under water at one side of the lower stand.

1 Choose two suitable glass cake stands and a single goblet-shaped glass for the top layer. Make sure that the top stand sits steadily on the one below.

2 Gently pour a little water from a jug so that it just covers the lower stand. Then do the same to the second one and fill the goblet.

4 Add a sprig of fuchsia alongside the lilies and put three pinks and a sprig of fuchsia on the opposite side of the upper stand.

6 Keep the water level topped up in the stands, so that the flower stems are always under water. You will find that water evaporates quickly from the large area, so keep an eye on it.

Beacon Rosa

Beverley

Rufus

5 Finally put several stems of fuchsias into the top glass filling it all the way round. Choose a small-flowered, dainty variety.

1 *Fill a shallow glass bowl with water and then gently scatter a handful of glass beads or marbles across the base.*

Beads, floating candles and fuchsia heads

Using fuchsia flowers upturned gives them an exotic look and by massing together several different varieties you can transform them into a stunning display. Fuchsia heads happily float in shallow water and if you combine them with glittering glass beads and flower candles in a round glass bowl, they make a party decoration for a sumptuous summer evening. Strong-colored varieties were deliberately chosen for the arrangement featured here to create the most impact, and the mixture of narrow-petalled types with rounder shapes gives a lovely lattice effect. Similarly, the decision to choose some very small flowers and mix them with large ones adds greater interest and impact than using blooms all of the same size. The candles are in two shades of pale pink and the beads are in a range of soft pearlized pinks. For a variation on this theme, you could combine some fuchsia flowers with loose petals from roses, or float other flowers and foliage among the fuchsias. It is always pleasing to include some flowers or foliage that have a fragrance to an arrangement like this.

2 *Next, place four or five small floating candles into the water. Be careful not to get water on the surface of the candles if you wish to light them later on.*

3 *Prepare a selection of fuchsia flowerheads by cutting the stems just below the point where they end in a small green lump.*

Cut the blooms neatly from their stems, using sharp scissors.

4 Once you have cut all the flower heads, begin to float them on the water. Arrange all the different types at random, but try to mix the various kinds as thoroughly as possible.

6 This display combines deep maroon and purple fuchsias with shocking pinks and reds. Salmon pinks and orange shades would make a different, but equally effective design.

Bob's Best

Delta's Wonder

Dollar Princess

5 Place the finished display on a low surface or in the center of a dining table for maximum impact. Ideally, the arrangement should be viewed from above.

If you light the candles, do not leave them unattended.

Peter Bielby

Baroness Van Dedem

The large furled petals of Cecille are shown off to great effect when upturned in this way.

Pink fuchsia candlesticks to adorn your table

Any arrangement or idea that exploits the fact that fuchsias hang their blooms downwards is bound to be successful. This idea uses just a few pieces of fuchsia or two different varieties but both basically pink. The leaves are also used as part of the whole arrangement and to help hide the foam ring, which is the crucial part of the idea. The candles can be lit or not, depending on how you wish to use the finished candlesticks. If you do light them, always stay in the room in case they burn down too far. Fuchsias are surprisingly good-tempered as cut flowers and should last well in a decoration such as this. After a day or so, check whether the foam is drying out and if it is, spray the ring with a fine mist or direct a little water straight onto the foam with a narrow spout. These white china candlesticks are particularly effective but you could use any design or color, or even plain, clear glass. Choose a medium-sized fuchsia flower that is quite light and dainty for a display such as this and avoid the very large, multipetaled types that would look wrong and top heavy.

1 Use a metal cookie cutter to cut rings from a piece of well-soaked flower foam about 0.8in(2cm) thick.

2 The candle makes the perfect cutter for the inner hole. Just push it down and through the foam. The candlesticks must have a large enough rim for the foam to sit on.

3 Put the candles firmly into place in the candlesticks and slide the rings over to sit on the tops of the candlesticks. They should be secure enough not to need tape.

4 Snip off small sprigs of fuchsia from larger stems. To achieve this effect, you will need one or two flowers, a few buds and perhaps some leaves on each little sprig.

Pink candles complement the fuchsias, but you could use a contrasting color or plain white.

6 Once there are fuchsias all round the ring, you can begin to fill in the spaces with smaller pieces and leaves to cover the foam ring around its top edge.

A few unopened buds look pretty amongst the greenery.

Do not discard the smaller leaves; they will be useful as fillers.

5 Begin to push the sprigs into the thickness of the foam, allowing the flowers to fall down naturally. Work evenly all round the candlestick.

7 The finished pair look pretty and just right to decorate a dinner table or to stand on a mantlepiece, where the flowers can be fully appreciated.

Index to Plants

Credits

The majority of the photographs featured in this book have been taken by Neil Sutherland and are © Colour Library Books. The publishers wish to thank the following photographers for providing additional photographs, credited here by page number and position on the page, i.e. (BL)Bottom left, (TR)Top right, (C)Center, etc.

A-Z Botanical Collection (Mrs. W. Monks): 27(BR), 55(Below TR)

Gillian Beckett: 55(TR)

Eric Crichton: 26(BL), 42(B), 79(TR,BR), 80(TR), 81(R)

John Glover: Half-title, contents, 10-11, 43(L),
75, 78, 79(L), 80(B), 81(TL), 82(BL), 83(TL,BL,R)

Clive Nichols Garden Photography: 81(BL), 82(BR)

Photos Horticultural Picture Library: 26-7(C), 27(R,CB)

Acknowledgments

The publishers would like to thank Pam Gubler and everyone at Little Brook Fuchsias for their help in providing plants and locations for photography.